Pi¡

MW01483163

Research
Needs

Proceedings of the Workshop

March 28-29, 1996
Held at the Lansdowne Resort and
Convention Center
Leesburg, Virginia

Organized by:
The Pipeline Research Committee,
Pipeline Division
American Society of Civil Engineers

Edited by:
John G. Bomba, P.E., F. ASCE
Kvaerner-RJ Brown, 1253 North Post
Oak Road, Houston, Texas, 77055

Published by
ASCE *American Society of Civil Engineers*
345 East 47th Street
New York, New York 10017-2398

Abstract:

Pipeline Research Needs is the proceedings of a workshop organized by The Pipeline Research Committee, Pipeline Division, American Society of Civil Engineers and held in Leesburg, Virginia, March 28-29, 1996. The purpose of the workshop was to identify and prioritize pipeline-related research needs. The workshop covered all pipeline applications, including crude oil, natural gas, water and sewer products, pneumatic, and capsule. However, instead of arranging the workshop along traditional industrial lines, the subjects were organized according to function. This integrated approach brought pipeliners from different industries together to share information and exchange ideas concerning such topics as: 1) Pipeline safety and protection; 2) design; 3) operations; 4) fluid mechanics/hydraulics; 5) construction and rehabilitation; 6) automatic control and instrumentation; and 7) freight pipelines.

Library of Congress Cataloging-in-Publication Data

Workshop on Pipeline Needs (1996 : Leesburg, Va.)
Pipeline research needs : proceedings of the Workshop on Pipeline Needs : March 28-29, 1996 : held at the Lansdowne Resort and Convention Center, Leesburg, Virginia / organized by the Pipeline Research Committee, Pipeline Division, American Society of Civil Engineers ; edited by John G. Bomba.
p.   cm.
Includes bibliographical references and index.
ISBN 0-7844-0246-9
1. Pipelines--Research--Congresses. I. Bomba, John G. II. American Society of Civil Engineers. Pipeline Research Committee. III. Title.
TJ930.W67 1996   97-16680
621.8'672--dc21   CIP

ACKNOWLEDGMENTS

The Pipeline Research Needs Workshop was organized by the Pipeline Research Committee of the American Society of Civil Engineers. The organizing committee of the workshop are grateful to the Sponsoring Agencies, Organizations and Societies for their contributions of time and money, without which this workshop would never have been held. We particularly want to thank Ms. Gretchen Hyde, for her untiring support and especially for her even temperment.

The funding received from the following organizations for the workshop is greatly appreciated.

- American Society of Civil Engineers;
- Office of Pipeline Safety of the U.S. Department of Transportation;
- Minerals Management Service of the U.S. Department of Interior;
- Structural Systems & Construction Processes, National Science Foundation;
- Office of Advanced Research, Federal Highway Administration;
- National Energy Board of Canada;
- Gas Research Institute;
- American Water Works Association Research Foundation;
- North American Society for Trenchless Technology.

TABLE OF CONTENTS

Keynote Address

Working Group Reports

Appendices

Terms of Reference
ASCE WORKSHOP ON PIPELINE RESEARCH NEEDS

I. PURPOSE

The purpose of the workshop is to identify and prioritize pipeline-related research needs. The workshop will cover all pipeline applications, including crude oil, natural gas, water and sewer, products, pneumatic, and capsule.

II. NEED

Buried pipelines are the most environmentally friendly and the safest means of freight transportation. Increasingly, the nation is relying on pipelines to deliver freight- in either liquid, gas, or solid form. The aging of major pipelines which were mostly installed following the economic growth of post World War II period has created a need to develop methods of assessing the remaining useful life, and of means of extending the life of these facilities. New areas into which pipelines are being built are constantly expanding. A large number of organizations are involved in sponsoring or conducting pipeline-related research activities. The proposed workshop will provide a forum to bring these groups together to identify and prioritize the major pipeline research needs.

An earlier workshop on buried pipeline research needs was held in 1987 at the University of Massachusetts. That workshop was sponsored by National Science Foundation. Since then, there have been major advances in the field of pipeline engineering. Pipelines are being installed in deeper and deeper water. There is a strong need to reexamine the recommendations of the 1987 workshop, and particularly, because of the strong onshore emphasis, examine offshore pipelining aspects in some detail.

The workshop participants will assess the state-of-the-art and develop a list of needed research topics on all aspects of pipelines. Such a list can best be developed through a workshop in which pipeline operators, researchers, planners, regulators, and engineers can get together and discuss research needs. It is also important that a forum be available for researchers in the water and sewer industry to communicate with the researchers in the oil and gas industry. This workshop will provide an opportunity to do so.

III. PARTICIPANTS

Participation is by invitation only. Fifty experts from the industry, government, and academia have been invited to participate. To encourage participation by knowledgeable pipeliners, an honorarium was offered to each participant.

IV. SCOPE

The pipeline industry has traditionally been segregated into oil and natural gas pipelines and water and sewer pipelines. In order to overcome this "segregation", this workshop will organize the subjects according to function such as design, construction, safety, operations, and automatic control, rather than by type of service. This integrated approach will bring pipeliners from different industries together to share information and exchange ideas. It will greatly enhance the dialogue among various pipeline groups that normally do not communicate with each other. More specifically, the program was organized under the seven general areas listed below:

(A) **Pipeline Safety and Protection**-- includes pigging for safety and integrity, pipeline leak detection and monitoring, pipeline spills, effects of earthquakes, hurricanes, and floods on pipeline safety, cathodic protection systems, and third-party damage prevention.

(B) **Pipeline Design**-- new design approaches for onshore and offshore pipelines (including earthquakes, flooding, etc.), possible revisions to design codes, and use of expert systems in design.

(C) **Pipeline Operations**-- any non-safety related operational issue such as use of drag-reducing additives to reduce power consumption, handling of emergencies and spills, economics of pipelines, pumping operation procedures, and maintenance of aging pipeline systems. Includes design for and use of pigs for pipeline cleaning, sizing, and entry ports, use of various instruments to detect pigs, leaks and corrosion. How to cope with hydrate formation and parafin build-up problems, particularly in deepwater oil and gas flowlines, will be explored.

(D) **Fluid Mechanics/Hydraulics of Pipelines**-- dynamic analysis of pipeline transients, water hammer and column separation, problems associated with multi-phase flow, cavitation in pumps and valves, rheology of slurry, and hydraulics of capsule flow.

(E) **Construction and Rehabilitation of Pipelines**-- new construction techniques for pipelines, and construction under extreme conditions such as cold regions, mountainous terrains, swamps and wetlands, and offshore conditions; in-situ lining, replacement of corroded pipe segments, retrofitting of existing pipelines to comply with earthquake design, renovating decommissioned oil pipelines and natural gas pipelines for other purposes such as transporting coal.

(F) **Automatic Control and Instrumentation** -- computers and other new technologies used for automatic control of pipelines, control strategies, and communication systems; includes measurement of flow, pressure, and temperature of the fluid in the pipe.

(G) **Freight Pipelines**-- slurry pipelines, pneumatic pipelines, capsule pipelines, and tube transportation systems.

V. TIME, LOCATION & DURATION

The workshop was held on March 28-29, 1996 , in Leesburg, VA, near Washington, DC.

2

VI. ORGANIZING COMMITTEE

The workshop was organized by:

Henry Liu, Ph.D., P.E., F. ASCE, Capsule Pipeline Research Center, Columbia, MO

William P. Quinn, P.E., F. ASCE, El Paso Natural Gas Company, El Paso, TX

John G. Bomba, P.E., F. ASCE, Kvaerner - R. J. Brown, Houston, TX

Ahmad Habibian, Ph.D., P.E., American Society of Civil Engineers, Reston, VA

Gretchen Hyde, American Society of Civil Engineers, Reston, VA

VII. REPORT

The findings of the workshop are documented in this report.

VIII. PARTICIPATING SPONSORS

The following organizations provided funding for the workshop:

- American Society of Civil Engineers (ASCE)
- Office of Pipeline Safety, U.S. Department of Transportation (OPS)
- Minerals Management Service/ Department of the Interior (MMS)
- National Energy Board of Canada (NEB)
- National Science Foundation (NSF)
- Gas Research Institute (GRI)
- American Water Works Association Research Foundation (AWWARF)
- North American Society for Trenchless Technology (NASTT)

3

IX. PROGRAM

The Workshop was organized and run according to the schedule included below:

Day One

8:00 - 8:30	Registration

8:30 - 9:45   Session 1:   Opening Session (Three Keynote Speakers)(Workshop chairman gave an introduction and guidelines on how the workshop will be conducted. Three keynote speakers presented general ideas on research needs for consideration by participants for discussion.)

9:45 - 10:00   Break (refreshments provided)

10:00 - 12:00 Session 2: Group Meetings
(Participants broke into 7 area groups. Each breakout group exchanged ideas on needed research in that group area.)

12:00 - 1:00   Luncheon (provided)

1:00 - 2:45   Session 3: Group Meetings
(Each group drew up a list of needed research topics in their area, and gave justifications for each.)

2:45 - 3:00   Break (refreshments provided)

3:00 - 4:45   Session 4: Writeup
(First draft of group report completed)

7:00 - 9:00   Dinner (provided; speaker-J Schrock)

Day Two

8:30 - 9:45   Session 5: Presentations
(Each group leader gave a presentation on the needed research in its area.)

9:45 - 10:00   Break (refreshments provided)

10:00 - 12:00   Session 6: Writeup(continued)
(Revision of first as second draft; turn in group reports to

workshop organizer.)

12:00 - 1:00   Luncheon (provided)

1:00 - 3:00   Session 7: (Wrap up)
                (Discussed the following questions: Where do we go from here?
                How do we prioritize the identified research needs? Where is the
                money coming from? How do we start?)

3:00   Adjourn

Summary:

Fifty recognized leaders in the water, sewer, oil, gas, and products pipeline industry were invited to participate in a first of its kind ASCE initiated workshop to identify what was needed to maintain the present pipeline infrastructure in an acceptable working condition, to discuss ways to improve the condition of the thousands of miles of deteriorating water, sewer, oil, natural gas, and products pipelines on which the people of the United States and Canada absolutely depend upon.

In addition, both newly perceived as well as long-standing pipeline related problems for which no good solutions, or at best, only partial solutions exist, were defined.

The challenges of pipelining in deep water were addressed. These included:
- high external pressure effects,
- the long suspended pipe length associated with surface installation techniques,
- other installation methods,
- the cold water temperature's impact on hydrates and pariffin formation inside flowlines,

Pressure surges due to valve closing (classic water hammer) have been extensively studied and control devices have been designed. The problem still exists. Similarly, multi-phase flow has been studied for years, mainly at reduced scales. Using the data on larger diameter pipelines is problematic. Slug predication, and the predication of the length of a slug by the most commercially available multi-phase flow software is very unreliable- mainly because there is not enough "real pipeline data" available to calibrate the computer models.

The generally poor condition of our aging water and sewer pipelines was discussed in detail. Potential methods/ways of improving these pipelines, and research activities designed to verify these methods were written up.

These problem areas are representative of those discussed in the workshop.

As funding specific research is a difficult endeavor, each group spent some time identifying governmental agencies and trade associations having a vested interest in a particular problem area, or more specifically, in the consequences of a real or perceived failure which might be prevented through research.

Finally, the cross fertilization of ideas generated by the diverse education and experience backgrounds of the participating personnel led to the identification of real world pipeline problems which needed solutions. The results of these varied discussions, including why research was needed and how it could be funded, are included for each of the seven groups as their specific research ideas.

Photographs of the Workshop

Henry Liu and Bill Hunt caught discussing the pros and cons of "Frieght" (coal and wood chips).

The last, last minute planning session - after breakfast - before the opening session, Workshop planners and the Group Moderators review last minute details

Tom Hoelsher, Transco, the Facilitator of Group C, Pipeline Operations, reviews the research needs identified by Group C.

Jim Baker, President, Baker Pipelines, his wife, and Dale Reid, Exxon Production Research Company, enjoying one of the fine lunches provided to the participants and their wives.

Welcoming Remarks
Henry Liu
Chairman, Organizing Committee

I wish to welcome you to this workshop on pipeline research needs, and to thank the sponsors of this workshop, including the following:
- The Pipeline Division of the American Society of Civil Engineers;
- Office of Pipeline Safety of the U.S. Department of Transportation;
- Minerals Management Service of the U.S. Department of Interior;
- Structural Systems & Construction Processes, National Science Foundation;
- Office of Advanced Research, Federal Highway Administration;
- National Energy Board of Canada;
- Gas Research Institute;
- American Water Works Association Research Foundation;
- North American Society for Trenchless Technology.

This workshop has exceeded the organizers' original expectations. It has attracted more sponsors, who pay a minimum of $5,000 (Some Paid $10,000) to make it happen. Initially, it was intended to be a national workshop. However, due to interest expressed by some Canadian Pipeliners, and the subsequent sponsorship by the National Energy Board of Canada, the workshop has gained an *international flavor*. Considering the close tie between Canada and the United States, and the many pipelines connected between the two nations, it is befitting and heartening that some Canadian experts are represented at the workshop.

The "pipeline" is a strategically important technology to any modern nation, especially the industrialized nations. Pipelines are used for transporting water, sewage, natural gas, crude oil, refined petroleum products, hazardous chemicals, carbon dioxide, coal, other minerals, grain and hundreds of other products. The importance of pipelines to the United States and Canada is no less than for any other mode of transportation except perhaps highways. Yet the public knows little about it because pipelines are mostly underground and hence, invisible--explainable by the proverb "out of sight-out of mind". The only time that the public is reminded of the existence of pipelines and realizes their value is during construction or when there is a problem such as a leak or rupture, which happens infrequently. Therefore, largely due to its obscurity, the "pipeline" does not enjoy the kind of attention and appreciation enjoyed by above-ground transportation modes such as railroad or highway. Yet, the pipeline is by far the safest and most environmentally friendly mode of freight transport, and, in many circumstances, the most economical way to transport not only oil and gas, but also solid freight. The public needs to understand that before strong public support for pipeline research is possible. This is a public education task that ASCE and other pipeline related organizations should undertake.

The purpose of this workshop is to identify and prioritize pipeline-related research needs.

9

This is not the first time that ASCE has attempted to identify and prioritize pipeline research need. In 1961, the Pipeline Research Committee of ASCE published research needs in the now defunct ASCE Journal of the Pipeline Division and as a special publication. The title was "Research Needs in Pipeline Engineering for the Decade 1966-1975." Then, in 1987, a workshop was held on "Buried Pipeline Research Need." Other meetings (non-ASCE) were also held to discuss research needs in specific areas, such as hurricane damage to offshore pipelines.

What set this workshop apart from previous events is the scope and diversity of the research areas addressed in this workshop, and the interdisciplinary approach. Instead of grouping the workshop along traditional industry lines according to natural gas, oil, water, sewer, and so forth, representatives from various industries were brought together to discuss research needs in safety, design, fluid mechanics and so forth. This approach maximizes cross-fertilization of ideas and avoids duplication in research. The subjects covered also extend far beyond the realm of civil engineering. I hope this workshop also will have a stronger impact than previous events, but this cannot be taken for granted. It will require follow up actions and continued hard work. To this end, I propose the following follow-up activities:

1.   Publication of the proceedings--this has already been planned for this workshop by the Steering Committee. Mr. John Bomba will serve as the Chief Editor. John is the Chairman of the Pipeline Research Committee of ASCE. The editorial board will consist of all the seven group leaders (facilitators).

2.   Encourage various agencies and organizations to develop or sponsor research in pipeline areas pertinent to their missions. Hope all the sponsors of this workshop will consider some research topics proposed at this workshop, and tap the expertise of those present here to develop more detailed research programs in specific areas.

3.   Consider possible legislation on selected areas that require government involvement, such as intercity transportation of future freight in large pipelines. Incidentally, the ISTEA (Internodal Surface Transportation Efficiency Act) legislation requires that the DOT Secretary report to the Congress on the economic feasibility of such futuristic pipeline systems. DOT and the Federal Highway Administration should take a lead on this matter.

4.   Organize a World Congress on Advancements in Pipeline Research and Technology. ASCE should take a lead on it, and should seek co-sponsorship by many other organizations so that it will be a truly interdisciplinary, international conference on pipelines. With three years of preparation, the Conference can be held in 1999 to welcome the arrival of the new century.

Finally, I wish to introduce the Steering Committee Members who organized this event: Mr. John Bomba who is chairman of the ASCE Pipeline Research Committee, Mr. Bill

Quinn who represents ASCE Pipeline Division Executive Committee, Dr. Ahmad Habibian who represents ASCE Technical Services, and Ms. Gretchen Hyde who represents ASCE Grants Department. Keynote speakers will be introduced separately. Again, welcome to the workshop!

KEYNOTE ADDRESSES

1. ***Pipeline Needs, Technology Opportunities for Natural Gas Transmission Systems***
   Theodore L. Willke
   GasResearch Institute (GRI)
   Chicago, Illinois

   Dr. Willke's Keynote Address was a narration based on the following slides. No written text is available.

Pipeline Needs, Technology Opportunities
for
Natural Gas Transmission Systems

Theodore L. Willke

ASCE Workshop on Pipeline Research Needs

March 28-29, 1996

Focus of Presentation

- Natural Gas Pipelines

- Transmission, Not Distribution Lines

- Broad Pipeline Needs

Topics

- **Trends and Issues**

- **Pipeline Needs**

- **Technology Opportunities**

- **Priorities at GRI**

Natural Gas Pipeline Statistics

- **Miles of Pipeline**
  - Transmission   196,000 Miles
  - Field   54,000 Miles
  - Storage   5,000 Miles
- **Total Compressor Stations**
  - Transmission   1300
  - Other   660
- **Pipeline Construction**
  - 1200 Miles
  - $650,000 per Mile

Source: Oil & Gas Journal, November 27, 1995

16

Trends & Issues

- **Little New Construction**
  - Looping of Existing Lines
  - Line Replacement
  - Increased Horsepower
- **Increasing Average Pipeline Age**
- **Competitive Market Pressures**
  - Aggressive Cost Cutting
  - Better Capacity Utilization
  - Increasing Demands of Customers
  - Abandonments or Line Conversions
- **Environment & Safety Pressures**
  - Ratcheting of Regulations
  - Public Desire for Zero Risk
- **Advances in Information Technologies**

Classes of Pipeline Needs

1. **Extension of Useful Pipeline Life**

2. **Prevention & Mitigation of Accidents**

3. **Increased Productivity (BTU/$)**

4. **Reduced Environmental Impacts**

Opportunities -- New Technology

- **Information & Communication Technologies**
- **Sensors & Measurement Systems**
- **Materials & Materials Systems**
- **Prime Movers & Compressors**
- **Gas Storage Technology**
- **Environmental Technology**
- **Risk Management**

Opportunities -- New Technology

Information & Communications Technologies

- **Pipeline Data Management Systems**
  - Enterprise Data Base Systems
  - Spatial Analysis Techniques
- **Real-Time Pipeline/Pipeline Optimization**
- **Real-Time Monitoring**
- **Global Positioning Systems (GPS)**
- **Cellular Digital Packet Data/PCS Communications**
- **Neural Networks/Fuzzy Logic**
- **Microprocessor Applications**

18

Opportunities -- New Technology

Sensors & Measurement Systems

- **In-Line Inspection/Smart Pigs**
- **Airborne Pipeline Integrity Monitoring**
- **Remote Sensing (satellite)**
- **Gas Measurement**
  - Advanced Meter Concepts
  - Energy Measurement on a Chip
- **Emissions Monitoring**
- **Leak Detection/Smart Mainline Valves**

Opportunities -- New Technology

Materials & Materials Systems

- **Ultra High Pressure Pipelines**
  - High Strength Steels
  - Welding Technology
- **Composites**
  - Repair Systems
  - Relining Systems
  - Composite Pipe
- **Advanced/Smart Coatings (Fiber Optics)**
- **Advanced CAD/CAE/CAM Systems**
  - e.g. Turbine Blade Design

Opportunities -- New Technology
Prime Movers & Compressors

- Low-Emission Turbine Combustors
- Better Engine Combustion
- Advanced Compressor Aerodynamics
- Higher Efficiency Gas Turbines
- Engine-Compressor Diagnostic Systems
- New Electric Prime Movers (Fuel Cells)

Opportunities -- New Technology
Gas Storage Technology

- High Value, High-Deliverability Storage Methods
  - Caverns
  - Peakshaving (LNG)
- Increased Flexibility from Existing Reservoir Storage
  - Higher Deliverability/Turnover
  - Integrated Optimization of Surface/Subsurface Operations
- Network Analysis for Pipeline-Storage Optimization

Opportunities -- New Technology

Environmental Technology

- **Emissions Measurement Systems**
- **Predictive Emissions Monitoring**
- **Right-of-Way Construction**
  - Remediation
  - Wetlands
  - Stream Crossings
- **Chemical-Biological Treatment of Contaminated Soils**
- **Active Noise Suppression Systems**

Opportunities -- New Technology

Risk Management

- **Risk Management Guidelines**
  - Program Design
  - Performance Measurement
  - Implementation by Pipeline Operators
- **Risk Assessment Tools and Models**
  - Estimation of Probability
  - Consequence Analysis
  - Risk Control Evaluation Tools
- **Pipeline Data Management Tools**
  - Mapping/Spatial Analysis
  - Work History Systems

Emissions Controls for Pipeline Compressors

- **Retrofit Kits for Reciprocating Compressors**

- **Low NOx Combustors for Gas Turbines**

Emissions Controls for Pipeline Compressors

- **Emissions Monitoring (PEM and CEM)**

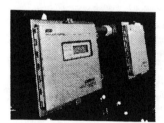

Tools for Pipeline Inspection

- Pipeline Simulation
  Facility (PSF)

- Improvements in
  Smart Pigs

Pipeline Safety and Risk Management

- Risk Management Tools

- Damage/Prevention and
  Pipeline Monitoring

- Automatic/Remote
  Mainline Valves

Pipeline Repair Using Fiberglass Composites

GRI

Engineering Guidelines for Uncased Pipeline Crossings

GRI

Improvements in Gas Flow Measurement

- Increased Field
  Measurement Accuracy

- Reduced Unaccounted-
  for Gas

Improvements in Gas Flow Measurement

- Low-Cost Energy
  (Btu) Measurement

- Low-Cost Electronic
  Flow Measurement

25

Increased Deliverability from Gas Storage

- **Reservoir Analysis (Simulation)**

- **Horizontal Well Technology**

Pipeline Right-of-Way

- **Erosion and Rehabilitation**

- **Stream Crossings and Wetlands**

GRI Priorities

- **Pipeline Inspection/NDE**
  - Crack Detection (SCC)
  - Detection of Mechanical Damage
  - Metal Loss Characterization
  - Unpiggable Lines
- **Integrated Spatial Analysis Techniques (ISAT)**
  - Pipeline Data Management
  - Computer Mapping
  - Digital (Data) Protocols
- **Risk Management**
  - Risk Management Implementation Guidelines
  - Incident Reporting And Trending System (IRATS)
  - Pipeline Inspection & Maintenance Optimization System

GRI Priorities

- **Composites for Pipeline Repair**
- **Emission Controls/Monitoring**
  - Low-Cost Emissions Monitoring
  - Predictive Emissions Monitoring
  - NOx Retrofit Controls for Compressors
  - Low-NOx Turbine Combustors
- **Gas Measurement**
  - Field Orifice/Turbine Meter Accuracy
  - Advanced Meter Concepts
  - Field Meter Proving
- **Site Remediation (Contaminated Soils)**
- **Storage Deliverability**

2.   **CRISIS: Funding for Infrastructure and Construction Research and Development**
      Thomas J. Pasko, Jr.
      Federal Highway Administration

ABOUT THE SPEAKER

Thomas J. Pasko, Jr. is the Director of the Office of Advanced Research in the Federal Highway Administration (FHWA). He started his career with the Pennsylvania Department of Transportation and then went to FHWA in 1961, where he conducted research in concrete and steel. He has been in management since 1976 and supervised research in many innovative concepts, such as epoxy-coated reinforcing bars, cathodic protection, modifications of and substitutes for asphalt, new deicers, and structural concepts such as pre-stressed pavements and high-strength bridges. He is a registered professional engineer, belongs to about 15 technical organizations, and serves on several advisory boards. He is presently the chairman of an interagency task group on infrastructure materials and a transportation committee on advanced materials. Under advanced concepts he funded a study of moving freight via powered capsules in tunnels. Mr. Pasko received a B.S. and M.S. in civil engineering from Pennsylvania State University and has completed two years of graduate work in transportation at Cornell University.

SUMMARY

The physical infrastructure and construction "community" is facing a crisis. The funding for Research and Development (R&D) in civil works has been woefully inadequate over many years and in comparison to other "higher" technologies such as weaponry, computers, energy, aircraft, communications, etc. The outlook in these budget-balancing times is that funding could be further decreased <u>unless</u> the civil engineering community raises concerns and speaks out to support the programs that improve the state-of-the-practice of civil infrastructure technologies. Government no longer has the resources to exhibit technology leadership in solving the problems associated with the fast-growing population needs and the demands for transportation.

THE PROBLEM

Table 1 below summarizes the dismal state of the construction industry's investment in R&D While the electronics industry invests over 5.5 percent, the construction industry is less than ½ percent, which is considerably less than the U.S. industry average of 3.8 percent.

Why is this so? Let's look at the following table.

TABLE 1 Industry R&D Investments	
Sector	Annual Investment (percent)a

Electrical/Electronics	5.5
Telecommunications	3.7
Aerospace	4.2
Chemicals	4.1
Automotive	4.0
Construction	0.5b
U.S. Average	3.8

a Business Week June 27, 1994. R&D Scoreboard. Business Week (3378): 81-103

b Civil Engineering Research Foundation. December 1993. A Nationwide Survey of Civil Engineering-Related R&D.

SOURCE: Civil Engineering Research Foundation. December 1994. National Construction Goals: A Construction Industry Perspective.

There are numerous reasons, some of which are mentioned in Reference 1. Among them are:

Fragmentation of the industry
- Use of ubiquitous, relatively low valued materials such as concrete and steel
- Lack of leadership and funding to effect market development
- Education levels of the workers
- Parochialism of the more then 89,000 organizations that own or control the country's infrastructure
- The conservatism of the civil engineering community to take risk (without adequate reward) and to persevere in the long time it takes to change codes, laws, and liabilities.

PRESENT R&D FUNDING

Of the $70 billion a year that the federal government spends for research, most of it is in the defense industry (Enclosure 1). Funding for transportation R&D is $700 million per year, and pipelines, which is the subject of this workshop falls under transportation. It is ironic that while transportation contributes about 20 percent to the Nation's Gross Domestic Product (GDP), it receives only about 1 percent of the Nation's R&D funding. Overall, there has been an inadequate amount of research funds for infrastructure and construction. The Department of Transportation (DOT) is primarily a regulatory agency. However, we do have almost $700 million a year for research, and it's hardly noticed out there in comparison to the other agencies.

When people talk research, they talk about the Department of Energy, the National Science Foundation, and the National Aeronautics and Space Administration (NASA). Very little is ever said about transportation. In comparison to the Transportation's efforts we can look at the Advanced Materials and Processing Program (AMPP) which was a 1993 program during a presidential emphasis on research and development (Reference 3). The Department of Energy devotes about $700 million of its total research budget of $6 billion a year to basic materials R&D, as contrasted to Transportation's total $700 million R&D budget. Many of the research dollars of AMPP are devoted to basic research into biological materials, fiber optics, superconductivity, etc. DOT is not heavily involved with basic research. For example, on the Maglev experimental electromagnetically propelled high-speed train, the superconductivity research is done by the Department of Energy, and on hypersonic aircraft, the research is done by NASA.

A lot of the Nation's research is either in defense or it is on the side of basic research and, to be a leader, you need lots of money in these areas. By comparison, the government and others have not invested a lot of money in construction R&D. Why has infrastructure not gotten the money?

FRAGMENTATION

Construction is a very fragmented industry, and many industries within construction, like steel and concrete, are not working together. The industry is comprised of many small firms that often have less then 10 employees, and most of the materials produced are of low unit value. For example, when you're putting down concrete pavement where you pay $20 per sq yard for a 10 inch thickness of concrete slip-formed into place, it costs less then the carpet in most living rooms.

The construction industry has not had an effective voice in the political scene. By comparison, the electronics industry has been effective. The composite industry is starting to do this, but the construction industry has not done it. Presently, the Civil Engineering Research Foundation (CERF) is attempting to pull together the construction community. A CONMAT (for construction materials) council has been created by 12 material sector representatives (concrete, steel, aluminum, composites, roofing, etc). They have put together a listing of $2 billion worth of additional R&D needs in order to improve the state-of-the-practice and enhance the USA as a global technology leader.

THINGS COULD BE WORSE

And, it appears that it may get worse. For instance the Corps of Engineer's Construction Productivity Advancement Research (CPAR) program was dropped from the Federal budget request in '96 in response to Congress. Hence, $6 million a year in government support is lost as well as about $6 million matching funds from industry, for a total loss of at least $12 million! The Corps literature sums it up this way:

"**Construction Productivity Advancement Research (CPAR-CRDA)** -- A cost-shared partnership to encourage technology transfer between the USACE laboratories and the U.S. construction industry. The objective of CPAR is to facilitate the application of innovative and advanced technologies through cooperative research and development, demonstration, commercialization, and technology transfer. To date the Corps has developed more than 72 partnerships under this program with a total investment of $67 million (Corps $26 million and industry $41 million). Congress ended this program in 1995."

Similarly, the Advanced Technology Program (ATP) of NIST (National Institute of Standards and Technology) is under attack and may not survive! The concrete industry had submitted a proposal to ATP for $160 million (to be matched by industry) worth of needed R&D. William Plenge writing in the September 95 Concrete International in "ATP-What Lies Ahead?" summed it up this way:

"What many consider the single, most important, most unifying initiative that the American concrete construction industry has undertaken since the advent of the interstate highway program in the 1950s now lies complete, pending approval but, alas, on the chopping block in Washington, D.C.—its funding held hostage by the new congressional majority taking aim at perceived "corporate welfare" in America.

Additionally, the Technology Reinvestment Program (TRP) of the Defense Department is also under attack. This program provided $22 million for the development of a Fiber Reinforced Plastics (FRP) composite bridge in California.

One needs to recognize that the concrete, steel, and aluminum industries have faced stiff world competition since the 1950s. In response, much of the past research capacity by industry has been severely curtailed. Whereas 40 years ago the Portland Cement Association (PCA) was funded by the cement producers to run a world-famous facility for a "cradle to grave" R&D program; today PCA spends little of its $2 million a year R&D money on basic research. Unbeknownst to most engineers today, most "basic research" on portland cement is government funded at $2 million per year through the National Science Foundation's (NSF) Advanced Cement Based Materials (ACBM) center at Northwestern University. There has been an almost complete flip-flop from industry funding to government funding. We are all challenged in the goal of global competition to bring government and industry together.

A similar situation exists in the steel industry where the Advanced Technology for Large Structural Systems (ATLSS) center at Lehigh University has a reputation for being the world-reknowned NSF Center for steel structural research, as well as other materials.

It is necessary to point out that although NSF has a new program focussing resources in the infrastructure technologies, NSF has not received additional funding to do more civil work. In addition to ACBM and ATLSS, there are only two other centers (in seismic and off shore technologies) in their total of 45 science and engineering centers. The others deal in the more exotic science (super conductivity, climate, atomic physics, etc).

31

Even now, other changes are taking place, it is rumored that the electrical power industry is getting apprehensive about the future of their Electrical Power Research Institute's (EPRI) program. EPRI funds about $300 million of R&D each year. A significant amount goes into infrastructure and materials R&D, in particular fly ash, and bottom ash as ingredients in concrete and controlled fills. Pending deregulation of the power industry may affect this industry R&D resource.

AND WHAT OF OUR FUTURE

I am now shifting away from our present needs in R&D for civil infrastructure and construction and want to look at the future. Today everyone is so busy "bailing the boat, that they do not have time to try to patch the hull, lest the boat sinks". Hence, our immediate needs are taking precedent over planning for the future. When I approach groups to consider future transportation, they respond with incremental changes to axle loads of trucks and modifications to the railroads. They fail to see that present congestion forebodes that in some areas our present transportation system is approaching imminent collapse. Pipelines, the subject of our workshop, could offer a partial solution to reducing the volumes of materials that presently move on highway and rails and provide some relief.

In looking at the future in Table 2, one of the trends is that cities are getting bigger. There isn't a city like Tokyo with a population of 27 million in the United States yet, and there may not be one by the year 2000, but sooner or later we are going to get there. For future large cities, I always think of the boulevards in Paris where about 8-story buildings just go on for miles. These buildings have been in place for more than 100 years, so you find a McDonald's restaurant housed in those type of buildings. In one center courtyard, a whole construction firm has all of its high-rise cranes. We haven't had to do this yet. We go out and we put a gas station in; we tear it down. We put up a McDonald's; we tear it down. We come in put up a bigger restaurant; we tear it down. Sooner or later we're going to have to conserve these buildings, and we're going to have multi-tenant housing, inadequate parking, and all of the congestion that goes with it.

What I would suggest, and this isn't the only approach one could use, is an improvement of the freight movement system. It could be put in place by tunneling under the city, like Japan is starting to do, and separating freight traffic from people. A 6-ft diameter pipe (2m) can be used to ship 4-ft modules. Self-powered capsules, similar to the old department store capsules that moved pneumatically, can be used to move freight. Other sizes of pipelines can be used to transport gasoline, milk, grain, coal, etc. to the centers of these large cities for processing and delivery to the consumers. What are the benefits of this?

In the Washington area, we have had numerous accidents and front page pictures of trucks rolling over on cars, or of gasoline truck disasters. This type of accident happens more frequently than we like. If you reduce the number of trucks on the road and move your freight in some other way, you get

32

FUTURE LARGE CITIES (MILLIONS)		
	1991	2000
1. Tokyo	27	30
2. Mexico City	21	28
3. Sao Paulo	19	25
4. Seoul	17	22
5. New York City	15	15
+		
+		
(13) LA	10	11
(27) Chicago	7	7

From the World Almanac 1994

just-in-time dependability. Our highway agencies are not going to change in their approach to avoiding traffic interruptions until the courts rule against them. In one state, the state had a construction job on the roadway and they shut down the traffic. One of the automotive plants shut down because its just-in-time delivery stopped. The plant went ahead and sent a claim to the state saying, "We can document what our losses are. We had to pay our labor. You owe us this amount of money." Well, it was dismissed because of state sovereignty. If and when one of those types of claims holds, then we will see a shift of the finanial burden from individuals (drives,businesses) to the government agencies. The agencies would then have to change their practices to provide more uninterruptable traffic flow.

Regardless, an underground system of freight movement is an all-weather operation, it lends itself to automation, is secure, and overcomes environmental concerns. It is going to get tougher and tougher in the next 20-30 years to build highways. We need to use advanced technology, and we have to innovate to resolve these issues in an evolutionary manner. More planning and R&D are needed to overcome the problems posed by the increasing populations of our cities before they become crises.

THE CHALLENGE

The barriers to innovation in construction infrastructure are many and diverse. Lack of funding, inadequate education, restrictive codes and laws, and lack of future planning are but a few. Identification of them is fairly easy, overcoming them will be much more difficult because of traditional and the organizational structures that protect the "barriers". Invention is easy! Changing the culture is difficult.

It is not adequate for us as civil engineers to spend two days in this workshop to produce a "Report on Research Needs". It is just another "Report" to add to the pile. We have to be proactive in raising concerns that investment in infrastructure R&D must be increased if the interests of the public are to be protected. We must start solving tomorrow's problems today through improved state of the practice. We can not wait until our system completely breaks down and the pressures from a rapidly increasing population makes it exceedingly more difficult to correct them in an incremental "band-aid" procedure.

Through ASCE's 120,000 members and your networks we need to raise the alarm and do our duty as "civil" engineers. Get involved!

REFERENCES

1.   Pasko, T. J., "Identifying the Barriers to Change," Federal Policies to Fast Innovation and Improvement in Constructed Facilities, Summary of a Symposium, Technical Report #129, Federal Facilities Council, National Academy Press, 2101 Constitution Avenue, NW, Washington, D.C. 20418 (Fax 202/334-3370) 1996. 150 pp.

2.   Civil Engineering Research Foundation "Materials for Tomorrow's Infrastructure" CERF Executive Report 94-5011-E. Civil Engineering Research Foundation, Washington, D.C. (Fax 202682-0612) December 1994.

3.   Anonymous, Advanced Materials and Processing, The FY 1994 Federal Program in materials Science and Technology; Civilian Industrial Technology Group on materials (COMAT) Task Group on National Planning, Chairman Lyle Schwartz, (Fax: 301/926-8849) July 1993.
     48 pp.

     Additional readings on related DOT and FHWA programs are:

4.   Brecher, A., Materials Research and Technology Initiatives, DOT-T-96-01, Research and Special Programs Administration USDOT, Washington, DC 20590 (Fax 202/366-3272), 1996. 51 pp.

34

5.   Anonymous, <u>Highway Research, Current Programs and Directions</u>, Special Report 244,
      Transportation Research Board, National Academy Press, 2101 constitution Avenue, NW,
      Washington, D.C. 20418 (Fax 202/334-3370), 1994. 112 pp.

**3.   *Pipeline Research Needs*
      John McCarthy
      Director, Engineering
      National Energy Board
      Calgary, Alberta, Canada**

Good Morning.

I would like to thank the organizers for providing me with the opportunity to participate
in the Workshop.

The National Energy Board and Research and Development

The NEB is an independent quasi-judicial tribunal responsible for federally regulated
energy matters. Briefly this means we are involved when energy crosses a provincial or
international border. Our purpose is to provide Canadians with fair, objective and
respected decisions on the variety of energy matters which fall under our jurisdiction.
Principally, we have four lines of business: we regulate the export and import of natural
gas and oil from Canada (similar to the U.S. Department of Energy), we regulate the tolls
and tariffs for interprovincial and international pipelines (similar to the Federal Energy
Regulatory Commission - FERC), thirdly, we regulate the exploration, development and
production of oil and gas from the Frontier areas of Canada (similar to the Minerals
Management Service), and, finally, we regulate the technical aspects of pipeline including
safety and environmental protection (similar to the U.S. Department of Transportation's
Office of Pipeline Safety). In fulfillment of these last two roles, we also have the
responsibility to manage a portion of the federal government's Program for Energy
Research and Development, known as PERD.

In total, our R&D program reflects about $2 million of disbursements annually to
independent contractors for the development of technology. Often these projects are
highly leveraged and we look for opportunities to participate in joint industry projects
which address a safety or environmental issue which is important to Canadians.

The Importance of the Canadian Pipeline Infrastructure

There are over 180,000 miles of pipeline in Canada including over 54,000 miles of large
diameter high pressure pipeline: 442,000 miles of natural gas and 12,000 miles of oil
trunk pipelines. Canadians rely upon our pipeline infrastructure. These pipelines convey

the energy necessary to heat our homes and feed our industry as well as provide significant export earnings. The energy sector represents 7 percent ($41 billion) of the Canadian GDP and oil and gas accounts for $8 billion of net exports.

Importance of Pipeline Integrity Management

Approximately half of our oil pipelines and 30% of our gas pipelines are over 30 years old. These lines, if well maintained and operated, should have an indefinite service life. Our economists suggest that the need for the pipeline infrastructure will be at current levels and beyond for at least another thirty years, considering the remaining reserves in the Western Canadian Sedimentary Basin and the potential additional resources from Western Canada, such as tar sands and coal seam gas. Obviously there is a strong economic benefit to maintaining these assets.

The Regulator's Challenge

Our regulatory role includes making regulations for the design, construction, operation and abandonment of pipelines and providing for the protection of property, the environment and the safety of the public and the company's employees.

From a regulator's perspective, the public seems to be demanding more than ever, that the pipelines which pass by their farms and homes are safe and reliable. This demand will certainly continue. In this world where the media influences public perceptions and political agendas, the inevitable failures which will occur will increase pressure on both the regulator and industry to regularly demonstrate that the systems are both safe and reliable and safe. To do our job properly, we have to be able to knowledgeably respond to these demands and ensure that pipelines meet an acceptable level of safety. In the current fiscal environment in which government operates, we must be creative in finding solutions.

Risk Management

Risk management offers the potential to enable both the industry and the regulators to meet this challenge. Canadian pipelines and regulatory agencies have started an initiative to encourage the staged implementation of risk assessment and risk management into the pipeline industry. This organization is called the Pipeline Risk Assessment Steering Committee (PRASC) and is similar to the RAQT in the U.S.

As time goes by, design and construction standards change, population centres grow and encroach upon the pipeline changing the potential consequences of an accident, perhaps the product being carried by the pipeline now differs from that at the time of design, over the years soil movement and third party activity may have introduced stresses and flaws into the system, perhaps the coating system has failed. All these represent possible and,

36

perhaps, expected changes. The operator will be challenged to demonstrate that in spite of these developments, the system still meets an acceptable level of reliability. Risk management provides a structured methodology to do this, but further work is required to refine risk management techniques and establish realistic performance measures.

Risk management requires data. Acceptable risk management requires acceptable data. One of the research needs often identified is a reliable and useable database. It is important to realize that the data is of equal value to the regulator and the industry. Given the variety of companies and regulatory agencies involved, it is important to ensure that terminology and data is consistent.

Surveillance and Monitoring Technology

Most of the pipelines that we should be concerned about haven't been seen in years. They are three feet under the ground in a unique micro-environment, and we can only infer how well they are performing by either a lack of visible failures or worse, obvious failures. Inspection technology is an important area of research. This includes the development of reliable in-line inspection tools and NDT inspection technology for identifying and sizing defects.

Effective surveillance of pipeline will be a challenge. Perhaps remote sensing technologies will offer some benefits, particularly, as companies operating in today's competitive environment reduce their field staff.

Understanding Time-Dependent Failure Mechanisms

As systems age, a variety of time dependent failure mechanisms will appear. Included in this list is corrosion fatigue, stress corrosion cracking, and external source hydrogen induced cracking. It is important to be able to understand these mechanisms and have available practical mitigative measures which would keep the pipeline in a safe and reliable state.

Repair Technology

Once defects are identified, it is important that the repair technology used remains effective for the remaining life on the pipeline and does not introduce another time dependent failure mechanism. Some of the word done with respect to the "Clock Spring" composite sleeves is very promising. Again, with these new technologies, it is important to be able to demonstrate the durability of the repair method.

Understanding the Performance of Systems

How well do we understand the reaction of buried pipelines to soil stresses caused by surface loads and ground movement? By understanding the interaction of these systems

37

we can understand and establish effective performance criteria. Limit states design methods should help.

Standards for New Systems

Lastly, it is important that we consider the design requirements for new systems to ensure that we benefit from our experiences. We must use the knowledge that is acquired from years of pipeline performance to ensure that these new systems will pass the test of time.

How Answers are Delivered can be as Important as the Answers

I have given a few thoughts of where research needs may be greatest. But I would like to suggest that involving several stakeholders in safety and pipeline integrity research programs can make these programs more effective.

In Canada, we have a long history of consensus standards in the pipeline industry. For example, all Canadian pipeline regulations, either federal or provincial, adopt by reference CSA standards. NEB staff actively participates on standards committees. We believe that this helps to build a strong technical understanding of the issues faced by industry. As well, reliance on the consensus standards ultimately reduces our enforcement costs as industry has a greater "ownership" in the regulatory requirements and is more likely to operate in compliance with these standards. At the same time, oversight by a regulator improves public confidence in the standards.

Similarly, participation by regulatory agencies and the public, where possible, in identifying research needs will help in ensuring that the technology meets the needs of all stakeholders and in ensuring that the technology which is developed is accepted.

In that light, I look forward to some good discussion of these topics over the next two days.

WORKING GROUP REPORTS

A.   Pipeline Safety and Protection
Facilitator: Tom Steinbauer, The Gas Research Institute

Group A reviewed the topics proposed by the workshop participants deemed to be relevent to pipeline safety and concluded that the most pressing need was the development of a real time tool for monitoring the integrity of a pipeline system.

1.   CONTROLS FOR FULL SCALE NATURAL GAS LINE BREAKS

Issue Definition:
A reliable, repeatable, practical method to remotely detect full scale natural gas transmission line breaks and close the valves isolation that failure has not been developed and/or successfully commercialized.

Background:
Although it had existed for many years, the natural gas transmission industry in the United Stated grew exponentially in the years following the second world war. With this growth, several pipeline companies became concerned with the detection of line failures (ruptures) and the automatic closing of isolation valves. These valves were spaced not more than 20 miles apart predicated on codes that the industry followed at that time - most notably those that evolved into what is known today as the ANSI/ASME B-31 code for Gas Transmission and Distribution Piping Systems - and still follows today, but under the requirements of 49CFR192.

Working with the then fledgling gas transmission industry, innovative entrepreneurs developed early versions detection devices of which the majority were based on rate of pressure decline. The rate of pressure decline principle was based on the theory that the speed of pressure fluctuations due to a line failure would exceed, by a sufficient margin, any pressure fluctuation due to normal pipeline operations. It took a number of years for the industry, who at the time did not have today's computer modeling capabilities, to learn that this principle was flawed.

A Gas Research Institute (GRI) Report entitled Remote and Automatic Main Line Valve Technology Assessment (GRI-95/0101) more fully described the various types of linebreak detectors and actuators that have been developed over the years. It also provides a rather complete analysis of the performance of rate-of-pressure-decline devices.

Justification of Needs:
Although natural gas transmission pipeline failures have been virtually eliminated in recent years, the potential for future failures due to third party will, for all practical purposes, never be completely eliminated. It is also a well accepted fact that the vast majority of damage that results from a gas pipeline failure occurs virtually at the instant of failure.

Given the unreliable performance history of line-break detection devices, the increasingly positive safety trends that natural gas industry is experiencing, and minimal benefit that the transmission industry perceives from line break controls, only a few companies have elected to install them.

Public perception is another matter. As in the case of the March 1993 Edison, New Jersey incident, local public officials claimed that a large percentage of the damage sustained as a result of that failure could have been prevented had line-break controls been installed and operating at the time of that failure. Thus, if public opinion is to ever be modified, and line-break controls accepted by the gas pipeline industry, a reliable, accurate and cost-effective line-break device must be developed.

Focus of Research:
Modern pipeline valves and operators have reached the point in their development that they are reliable. Thus no further research is immediately apparent for these components in this application.

What is lacking is an understanding of the first-principles of signal that would reliably identify a line break. While the above referenced report indicates that Acoustic Systems, Inc. (ASI) has refined the earlier "wave alert acoustic development of Dr. Morris Covington, the efficacy of this technology needs to be conformed a "basic research" exercise, in addition, competing technologies must be developed and, along with the ASI technology, advanced to viable commercialization. Thus a comprehensive effort, starting with a basic understanding of the signals indicative of line breaks, and carried through commercialization, must be pursued if line break controls are to become a reality.

Potential Sources of Funding:
Potential sources of funding for this development are as follows:
* Pipeline valve and pipeline valve control manufacturers
* Interstate Natural Gas Association of America Foundation (The INGAA Foundation)
* Pipeline Research Committee International (PRC International)
* Gas Research Institute
* The Department of Energy
* The Department of Transportation

2. REAL-TIME PIPE INTEGRITY MONITORING RESEARCH NEEDS - Jim Liou

Issue Definition
To monitor a pipeline continuously over time for signs of breach or degradation of integrity. Such signs may include leaks, dents, wall thickness reduction, and cracks.

Background
Pipelines have long been recognized as the safest mode of long distance high volume transport of crude oil, and natural gas. Despite this fact, pipelines can and do fail. General public's low tolerance for pipeline mishaps has resulted in strict regulations and potentially high fines that make leaks very costly. Aside from the financial concerns, leaks may cause environmental hazards and

41

threaten public safety. As a result, pipelines integrity is a key issue in many aspects (such as performance, safety, life extension, and risk assessment) of pipeline operations.

To prevent failures, pipelines are subjected to rigorous testing before and during installation. Corrosion-resistant waterproof coating and cathodic protections are routinely sued to deter corrosion during the service life span. However, due to wide variations in environmental variables, these protective measures are not completely effective and defects still can and do develop over time. Furthermore, some failure mechanisms are not well understood, let alone their prevention. Thus, assessment of pipeline integrity is necessary.

Justification of Needs

Integrity assessment is being done by means ranging from serial surveys (low cost, non-disruptive, but of limited value) so hydrostatic testing (high cost, disruptive, but comprehensive). Among these, intelligent pigs emerge as the method of choice for some oil and gas pipelines. However, intelligent pigs are expensive to run (several thousand dollars per mile) and the results may not be reliable. Under best circumstances, intelligent pigs only provide data for infrequent periodic integrity assessment (once every two to ten years, for example). The high cost prohibits more frequent inspections yet defects may develop and grow between inspections.

Furthermore, to use pigs, pipelines must allow free passage and must have pig launch and retrieval facilities. At the present, 42 percent of natural gas pipelines, 11 percent of oil pipelines, and all water mains in the United States cannot handle pigs due to physical limitations.

In terms of capability and cost, there is a gap in technology and tools for pipeline integrity assessment.

Focus of Research

1.   Assess the industry's interest in real-time non-intrusive pipeline integrity assessment.
2.   Establish performance requirements.
3.   Identify applicable concepts and technologies.
4.   Evaluate the technical feasibility of each.
5.   Rank the suitability-to-task of each approach.
6.   Recommend the most suitable and promising concepts and technologies for further research and development.

Funding Sources

- National Science Foundation
- American Petroleum Institute
- Gas Research Institute
- Private Investors
- Operation Pipeline Operators (for Prototype tests)

B. Pipeline Design
Facilitator: Raymond Sterling, Trenchless Technology Center

The members of Group B evaluated all of the pipeline design related proposal submitted by workshop participants. After considerable discussion, three major pipeline design research needs were identified. These are discussed below.

1. DESIGN GUIDELINES FOR TRENCHLESS TECHNOLOGIES IN NEW CONSTRUCTION

Justification:
Trenchless technologies are divided into three areas of application:
1.   Rehabilitation of existing pipelines;
2.   Methods of pipeline replacement of existing pipelines;
3.   Methods used for installing new pipelines.

Design related to Rehabilitation is covered in another section. Items 2 and 3 are in fact total replacement of existing pipelines in place with new pipe or the installation of new pipe where no pipe existed before.

In light of the use and application of trenchless technologies to accomplish these tasks new design considerations must be made. These new considerations will require the formulation of new standards for material and new standards for material handling and the process of installation.

In addition to operational consideration the use and application of new or emerging technologies will require study and development of application information to assist the end user in determining the appropriate technology to use for a specific application. The study should address areas such as environmental as well as social-economic impact, safety, speed of installation, cost and system operation.

Focus of Research:
Design methodologies for pipelines are dependent on the installation procedures. Issues that must be considered include initial stresses caused by installation, soil support conditions, and loadings, both external and internal. Many of the new trenchless technologies differ greatly with respect to installation procedures, resulting in a range of possible pipe conditions during and after construction. The primary emphasis of research in developing practical guidelines for trenchless technologies would be directed towards broad standards of practice that would not limit the development and application of proven and emerging technologies, but would assist the pipeline owners in deciding what methods are most applicable, what limitations might govern the various methods, what pipeline materials are suitable, and what minimum standards of practice should be exercised to assure a safe and reliable pipeline system.

43

The tasks required for this research are outline below. While not all tasks are strictly research topics, they are necessary to generate a useful set of design/installation guidelines.

Task 1   Assess the State-of-Practice for Trenchless Technology.
The main emphasis for this task is to identify broad characteristics of the various trenchless technologies. Much of this work has been accomplished n a variety of technical and trade publications. The goal would be to summarize broadly the generic types of methods and identify the range of pipeline geometries and installation limitations associated with the technologies. This task would consolidate accumulated industrial experience, and would serve as a preliminary filter for utility engineers faced with decisions as to acceptable methodologies. This task also would help when deciding the applicability of new techniques to particular needs.

Task 2:   Design Procedures for Evaluating Pipelines Installed using Trenchless Technologies.
Pipelines are designed to withstand external and internal loadings. Design criteria typically are based on limiting stresses or deformations, along with considerations for long-term survivability. Differing trenchless installation methods can impose initial stresses in pipelines as a result of pulling forces and geometric effects such as curvature resulting from changes in elevation and horizontal offset. Limits to acceptable pulling forces, pipe elongation, and geometry variations need to be identified. These limits necessarily will depend on the pipe materials and sizes used, and the joining methods for the pipe sections. Simplified guidelines for acceptable installation limits would provide pipeline owners greater confidence that the installation procedures do not affect the integrity of the final product. Also, tolerable installation limits would assist contractors in selecting the most cost-effective installation procedures acceptable for a particular pipeline size and material.

Task 3   Stress-Based Design Guidelines.
The goal of this task is to develop a general design methodology for buried pipelines that recognizes the interactions between pipe installation procedures, contract conditions between soil and pipe, and the interactions between the soil-pipe system. The deliverable from this task would be a methodology that identifies the stresses or deflections that develop certain classes of trenchless installations, identifies the limiting or allowable stresses or deformations for pipelines of various materials, and allows the owner or designer of the pipeline to match the pipeline material with the function and stress environment. This task requires an understanding of the mechanical behavior of pipelines installed using trenchless technologies, as well as the limit states for various pipeline materials and joining methods. The ultimate goal of this task would be a method by which a pipeline owner could work more effectively with the developers of trenchless installation technologies to decide on the suitability of different pipeline materials for differing

44

conditions.

Suggested Funding Resources:
The funding for the improvements in design methodologies should be shared among industry trade associations for the utility industries (e.g. AGA, GRI, AWWA, WEF, etc.; cities and communities facing major upgrading efforts; and the federal government agencies (e.g. U.S. DOT). The research and technology transfer efforts would be carried out by combinations of university efforts, joint industry efforts, proprietary research and collaborative activities among these.

Issues:
Regulatory acceptance of new trenchless technologies, materials and design techniques is an important implementation problem. Efforts similar to the HITEC program (Highway Innovative Technology Evaluation Committee) within ASCE can assist in this regard.

2.   LIMIT STATE/RELIABILITY-BASED PIPELINE DESIGN

Introduction:
This topic refers to a design methodology for pipeline engineering and pipeline design. The process involves identifying all failure modes leading to a limit state. Distinctions are made between those that imply failure (such as rupture) and those which imply an impairment of serviceability (such as excessive ovalization that makes it impossible for an intelligent pig to move along the line.) Limit states can be assigned statistically meaning by using probabilistic or stochastic weighing of inputs, such as loads, material behaviour. Therefore, the design methodology can be extended from so called "limit-state" analysis to "reliability-based design", where a probability of failure can be calculated fro a specif design.

Justification:
Existing design codes (e.g. B31.X, CSA 662) are largely based on stress limits and satisfactory experience with these limits in previous designs. However, the pipeline industry continue to pursue new designs that deviate significantly from past experience in one or more of the following areas:

- New materials (e.g. higher strength steels)
- Higher Loads (e.g. Collapse loading in deeper water, ice loading in arctic)
- New construction techniques (e.g. High residual strain, stress)
- Unusual service conditions (e.g. High temperature, high fatigue)

Existing codes, in many cases, can not adequately address these conditions while maintaining levels of safety consistent with more conventional design. Therefore, there is incentive to pursue new design methods that will result in:

1)   an improved understanding of risks;
2)   consistent levels of safety;
3)   more economical design;
4)   better understanding of design behaviour;
5)   an ability to rationally address new materials and technologies.

There is significant effort ongoing in this area in Europe and Canada. The upcoming ISO 9000 Standards are likely going to reflect these efforts to some extent. By comparison there is extremely little effort underway in the US. As a result, US authorities, industry may be unprepared to understand, accept, or challenge forthcoming standards. In summary there is a need to better understand, and develop improved design and engineering methods based on limit state and reliability based design, and be prepared to intelligently contribute to the improvement of the US, Canadian and international design standards.

Focus of Research:
The principal steps proposed in this area reflect a practical and systematic approach to the

46

development of new design criteria and methods:

1) Review, consolidate and monitor existing research and development activities and available products, such as those that have been or are being carried out in Norway (e.g. SINTEF program), Netherlands (new design code), Denmark, and Canada.
2) Identify potential strengths, weaknesses, and "holes" in this body of work.
3) Identify the range of applications of interest to the pipeline industry (assuming most applications would likely respond favourably to such an effort).
4) The principal body of work would be in developing analytical models that address the range of applications failure modes, limit states of interest. Several of these may have already been developed.
5) Conduct the necessary level of testing and calibration of these models to laboratory data, industry data, and experience.
6) Pursue incorporation of limit state design methods, based on above, into the relevant design codes.
7) Seek out and consolidate probabilistic data that could be used to support development of reliability based design tools, extending from the limit state design methods already developed.
8) Develop reliability based design methods and criteria.
9) Encourage development of practical, cost-effective design tools and software for general engineering use.
10) Provide a mechanism for sharing and communicating new advancements, and industry experience within US, Canada and international pipeline community.

Funding Resources:
The proposed effort is very large, the necessary involvement of affected parties is broad, and the financial resources required would be significant (on order of 10s of M$).

Potential funding sources would likely include:

- Industry trade organizations (e.g. GRI, API, others).
- Federal Government (DOT, MMS).

Participation:
Much of the work could be performed through the coordinated involvement of commercial research organizations, university research groups, along with participation by organizations such as ASME, ASCE, and IEEE.

Issues:
Two leading issues affecting the successful pursuit of the effort which:

1) Coordination of all assigned efforts through a central organization or committee.
2) Participation of all affected parties including industry, government, academia, and

international standards organizations.

3. RAPID JOINING

Justification:
Rapid pipe joining refers to new welding processes and/or mechanical joining methods that will permit faster offshore pipe joining as compared to conventional welding. The offshore industry is continuing to move to deeper water which may require pipelines to be exclusively laid using a method called "Jlay" where pipe is stacked and welded in a near-vertical orientation where all pipe joint welding, inspection and coating are done at a single station. Because of this, productivity and pipelaying cost are determined by the speed of these operations. If rapid jointing could be made available, with performance and reliability comparable to conventional welding, a 25-40% savings on installed cost may be realized. Additionally, rapid mechanical joining methods would enable the use of low-cost, non-weldable corrosion resistant alloys such as Cr13 ("Chrome-13") pipe, resulting in even greater cost savings. Resulting shortened laying schedules will also result in less environmental exposure and faster project completion. The development of rapid joining methods should also generate greater interest and make deep water pipelaying more attractive.

Focus of Research:
The future focus for the development of rapid pipe joining technology should be centered in four major areas:
- Development of new rapid welding process, e.g. homopolar welding, laser fusion, and others.
- Development of mechanical connectors for metal line pipe, i.e. threaded connectors and others.
- Development of new and more reliable joining processes for pipes composed of plastic and composite materials.
- Development of reliable inspection techniques to certify joint integrity for the above processes.

Developments within these areas should concentrate on intiating new or improving existing processes, conducting prototypes, intallations, and shaping NDE techniques to verify the integrity of rigid joining techniques. Research should be initated to include the application of rapid joining processes to traditional materials and testing techniques applicable to both onshore and offshore pipelines, installations, but also to the more extreme applications associated with deepwater developments offshore requiring high-strength corrosion steel alloys. Due to the large cost savings that may be associated with the use of such techniques in new pipeline installating, the intent is develop rapid welding processes and mechanical connectors such that the comparable performance and reliability equals or exceeds that of conventional welded or mechanical joints.

Suggested Funding Resources:
The development of improved and rapid joining systems should be funded by suppliers of the pipe and/or joint systems. This funding should be leveraged by leveraged by industry operators and the government (DOE, OPS, MMS) based on their respective interest in faster, cheaper and more reliable joining techniques. The work would involve university research on breakthrough

techniques, commercial R&D by suppliers on improvements to techniques, and contractor involvement in development and testing. Government funding is needed for breakthrough techniques and may be desirable for evaluation. International cooperative funding may be possible for some joining applications.

C.   Pipeline Operations
   Facilitator: Thomas Hoelscher, Technical Manager Field Operations, Division III,
   Transco

The Pipeline Operations Group members represented the oil, gas, water and sewer sectors of the pipeline industry, and included engineers from construction contractors, design engineering firms and operating companies. The 30 plus ideas presented by workshop participants, as well as those developed within the workshop, were carefully reviewed. The surviving ideas worthy of further consideration and possible research are discussed below.

1.   ASSESSING THE NEED FOR DEEP WATER PIPELINE REPAIRS

Justification:

Offshore production, currently ongoing at 5,000 foot depths, will go into deeper waters and may require sub-sea pipelines. Damage to pipelines, whether leaks or mechanical damage only, can be caused by mud slides, collapse, installation accidents, and corrosion. The technology for repairing pipelines at these depths does not exist at this time. Should damage occur, the only option available to continue production is to replace the damaged pipeline with a new pipeline. This replacement pipeline is so expensive and that it could cause abandonment of the producing field.

Focus of Research:

A preliminary study would include collecting information on the following:

- The location, depth, service and diameter of pipelines that exist and or planned that are below the depth at which current repair practices are applicable. An estimate of the frequency repairs of should be made;
- The state of the art sub-sea repair techniques and the possibility of using these techniques at deeper depths. This would include a listing of vendors, production companies and contractors with pipeline installation and repair experience;
- The cost of developing a deep water pipeline repair system including tools, hardware, service equipment, storage and maintenance;
- The cost of making a repair to a sub-sea pipeline as opposed to replacing the pipeline.

Funding:

Funding for the study should be provided by the following:

- Sub-sea production companies
- Vendors
- Offshore pipeline contractors
- Governments currently affected by deep sub-sea production

Follow Up:

Develop a cost benefit analysis comparing the cost of pipeline repair vs. replacement.

2.   PREVENT PIPELINE BLOCKAGE DUE TO HYDROCARBON SOLIDIFICATION.

Background:
Following the advent of Federal Energy Regulatory Commission (FERC) Order 636 and the subsequent trend towards transportation of saturated natural gas and condensates, hydrates formation has become a significant problem for offshore operators. At the same time production has been moving to greater depths in the Gulf of Mexico. Here the conditions of free water, condensates, higher pressures and lower temperatures that are conducive to the formation of hydrates in pipelines are frequently encountered. It is not always possible to recognize hydrate formation conditions, to prevent their solidification and to safely disperse them once formed.

Justification:
Further research in this area will enable operators to improve safety and prevent damage to pipelines and property. Clearing hydrate formations using present technology, i.e. reducing the pressure in equal increments on both sides of the hydrate, can cost up to $500,000 per incident. A secondary benefit of the research will be a reduction of the loss of throughput and revenue to production and pipeline companies.

Focus:
The dual focus of research is the prevention of hydrate formation and development of a cost effective method of dispersing the hydrates should they form. Research on possible methods of dispersal could include development of technology to introduce heat to the hydrate formation from the exterior of the pipe. Remotely operated vehicles or use of steam lines are possible concepts to investigate. Other avenues include development of new hydroscopic materials to disperse hydrates. Consideration of hydrate formation during pipeline design also needs additional emphasis to determine if insulated lines, heat tracing, high efficiency internal coating etc may be an effective means of prevention. Other hydrocarbon solids, e.g. paraffines and asphaltines, can also cause blockages of pipelines. This research could benefit those problems also.

Funding:
Potential sources of funding for this work could include the offshore producers, pipeline operators and governments.

Follow up:
- Risk analysis;
- Probability analysis.

3.   DEVELOP A COST EFFECTIVE AND NON-INTRUSIVE PIPELINE INSPECTION METHOD

Justification:

Pipelines are subject to many forces that can cause a loss of integrity. Third party construction activities can cause dents and gouges. Missing or disbonded coating can allow general as well as microbiological corrosion. Defective welds, both girth and longitudinal, as well as other manufacturing defects can cause leaks and failure.

Pipe coating and line pipe integrity are currently determined by infrequent periodic inspections. Pipe coating is inspected by impressing a current on the pipe and analyzing its leakage through coating defects along the pipeline. Offshore coating surveys are very expensive. Tape coating can become disbonded from the pipe which can not be detected with technology available today. Hydrostatic testing can affirm pipeline integrity, but is costly and requires pipeline outages. As a result, the percentage of pipelines hydrotested on a regular basis is extremely small.

Smart pigs are available that can detect metal loss and other pipe anomalies, but require downtime, loss of throughput and modification of existing facilities - all of which are very expensive. The smart pig logs must be interpreted before judgements are made on the physical cause and severity of each signature. Only then can subjective risk management decisions be made on which anomalies should be physically inspected. Many pipelines are offshore or under rivers and roads, making physical inspection difficult and costly.

Many pipelines cannot be smart pigged today. Reduced size valves are a challenge and plug valves completely prevent pigging. Even in lines that are piggable, the cost to modify facilities to physically allow pig runs has a very high capital cost as well as a loss of throughput. In addition to these costs, the smart pig runs alone range from $1,000 per mile to $5,000 per mile and more.

Focus of Research:

At its best, the use of smart pigs can accurately and reliably detect, size and locate the various types of defects that can exist or be inflicted on a pipeline. Research is needed to be able to obtain data comparable to that obtained with smart pigs in pipelines that cannot be pigged. This should be done in a manner that does not damage the line pipe or the coating. The options are to:

1. detect defects remotely, either from the ground or the air over the line;
2. detect defects internally, by means of a mini-pig that can readily traverse the constrictions and bends that make lines unpiggable; and,
3. detect defects from perturbed signals in the pipe wall. Combinations of two or all three might also be considered.

The elements of the research needed to develop a non-intrusive, cost effective, broadly applicable inspection system should include:

- Methods to excite the pipe wall
- Methods to detect the signals at positions remote from their generation point

- Methods to delineate the presence of anomalies from these signals
- Methods to distinguish individual types and sizes of damage from signals showing the presence of anomalies.

The research should be divided into new construction and existing pipelines. In new construction, many straight-forward opportunities exist to make the pipe "self-examining" before it is placed in service. However, as the opportunities to modify existing pipelines are very limited, research will require a quantum leap from the currently available technologies. Workshops and other synergistic activities should be undertaken that will match a high level of technology (eg at the National laboratories) with the reality of the practical service conditions in which pipelines operate.

Funding Source:
Pipeline Companies & Associations
Vendors

D.   Fluid Mechanics/Hydraulics of Pipelines
Facilitator: E. B. Wylie, Department of Civil and Environmental Engineering, University of Michigan

The Fluid Mechanics/Hydraulics Group, which included representatives of academia, research organizations, consulting engineers, and gas and electric utilities, recommends the following four areas for further investigation and or research. These areas encompass water quality, long term degradation of interior pipe wall, and transient phenomena in both the water and the oil and gas industry. Both onshore and offshore pipelines are included.

1.   WATER QUALITY ISSUES IN WATER DISTRIBUTION SYSTEMS

Justification:

Over the last decade, because of changes in the Safe Drinking Water Act, the understanding of water quality, and the need for modelling the quality between the treatment plant and the customers' taps has become increasingly more important. The EPA and organizations such as AWWA have done much work in broadening the understanding of the various factors contributing to the overall quality of water in the distribution system. The existing models do a reasonably good job in predicting "conservative" constituents such as fluoride and salinity and substances that exhibit first order decay.

There is still a great deal of work that needs to be done in modelling the behavior of more complicated substances/characteristics in/of water such as pH-alkalinity-hardness disinfection byproducts, or sediment transport. In many instances, reactions that occur at a given rate in laboratory vessels do not occur at the same rate in the water distribution pipes. This is probably due to the reactions with pipe walls and any biofilms present. These reactions and their rates are not well enough understood. Basic research is needed in some of these areas.

In addition, there also is a need for integrating the existing work in this area performed by EPA, AWWA, and others, setting up a theoretical model containing as many of the variables as practical and following this up with calibration on real systems (parts of existing, operating water distribution systems), thus proving or disproving the theoretical model.

Focus:

In view of the overall public concern of water quality, we must define companies/agencies with common interests to leverage funds to achieve a working calibrated water distribution system prediction model.

Sources of Funding:
EPA and AWWA

2.   LONG TERM HYDRAULIC EFFECTIVENESS OF REHABILITATED PIPELINES

Justification:

Because of the aging infrastructure, rehabilitation of existing pipelines is economically attractive. This rehabilitation may be achieved by inserting a smaller diameter pipe, constructing liners, or by cleaning. The rehabilitation design engineer must be able to accurately estimate the flow quantity which can be handled by the "rehabilitated" pipeline and to predict with some certainty, how long a time will the system work efficiently--before it needs to be re-rehabilitated. Better prediction ability would result ins sounder economic decisions. Research is needed which will provide data for accurately forecasting flow capacity for various types of rehabilitation options. This is achievable by performing an expost analysis of roughness factors ("C" factors, equivalent sand grain roughness, etc.) immediately after rehabilitation and at periodic intervals, say 5 years, to enable one to predict long term changes. In other words, what is the life of a rehab project?

Focus:
Pitometer Associates, and like organizations, among other sources, maintains a data base of sorts which documents some of the rehabilitation projects. With the assistance of these organizations, assemble available data, analyze, and establish a predictive model.

Sources of Funding:
AWWA, EPA

3.   UNSTEADY FLOW IN PIPELINE SYSTEMS

Justification:
Although a fundamental understanding of transient flow in pipelines exists, accidents, on occasion, plague some facet of most of our industries - water, sewage, oil, gas, and product pipelines. This is probably because the topic is not addressed until after an event.

However, some hydraulic issues remain, primarily where field data are lacking. One area deals with check valves, where data only exist from laboratory scale experiments. These data involve the dynamic forces during forward and reverse flows for various valve openings, and they form the basis for any theory and modelling capability that exists. Experiments are needed on larger valves, since size scaling does not work too well. Current modeling techniques do not necessarily conservatively predict the performance of prototype systems.

Other unsteady issues include liquid vaporization and condensation, unsteady multi-phase and multi-component flows, transients induced by thermal effects and air travel and removal in liquid lines.

Focus of Research:
Gather prototype size transient data to validate numerical models.

Source of Funding:

NSF, AWWA, EPRI

4.   CALIBRATION OF PREDICTION TECHNIQUES FOR SLUG FLOW AND SLUG LENGTH

Justification:
Slugs due to pigging and terrain features are predictable within plus or minus 20 percent. However, currently available prediction techniques for hydrodynamic slugs are particularly bad. Similarly, prediction techniques for determinintg slug lengths are worse.   One commercially available program has a reported (by its representative) of plus or minus 100 percent.   The next best has a reported accuracy of prediction of plus or minus 1000 percent.

Both of these reported accuracies may be exaggerations to some extent, but basically are true.

What is required is field data for operating multi-phase pipelines to calibrate the prediction techniques for available multi-phase flow programs.   These data would be used to derive new, or update existing correlations for slug flow and slug length prediction.

Focus:
Approach companies with operating multi-phase flow pipelines (mostly Gulf of Mexico) for available data, and where necessary, to ask them to record additional data.   These data would be used to adjust existing correlations to fit new data.   These new correlations would be published.

SOURCE OF FUNDING
IGT, SGA, API, Companies with multi-phase programs as well as those with multi-phase flow problems.

E.   Construction and Rehabilitation of Pipelines
     Facilitator: Jay Schrock, JSC International Engineering

Group "E", which developed the research needs statements in the area of pipeline repair and rehabilitation, included consulting engineers, user/agency representatives, pipeline contractors, a plastic pipe materials supplier, and the ASCE Pipeline Division's Technical Liason representative. The Group represented an extensive background in the rapidly growing area of pipeline repair and rehabilitation.

1.   DEGRADATION IMPACT ON THE STRUCTURAL ADEQUACY OF PIPELINES AND DEGRADATION CONTROL

Introduction:
A common practice is to rehabilitate several thousands of feet of pipeline, when in fact the length that is structurally impaired is a small fraction of the total length. Rehabilitation costs run in the range of from five to twenty dollars per inch of pipe diameter per linear foot, so the cost for a five foot diameter pipe can run about $1.5 to $6 Million per mile.

Several factors lead to this over conservatism among rehabilitation designers, not the least of which is the liability issue. since in many contracts the designer is required to "hold harmless and defend" the agency for any and all future problems, regardless of fault. This is a case where the "risk management" attorneys on the owner/agency's staff increase the expenditure of private funds and tax dollars by factors of five to ten or more, thinking they are "protecting" the agency/owner/client. The second factor, and one which requires some testing and research, is the need for verification of design methods which are capable of predicting the load carrying ability of a deteriorated pipe. The fact that the pipe, prior to rehabilitation, is carrying soil and live loads and internal pressures successfully is certainly evidence that the corroded pipe possesses significant structural capability, or that the present loads are less than the original loads.

Justification:
●   We must develop the ability to predict failures before they occur, to protect the public and the environment;
●   We need to develop mitigation measures to reduce property damage and loss of life due to pipeline failure;
●   Methods are needed to accurately assess the remaining life of a deteriorated pipeline, so that we can properly allocate scarce rehabilitation resources by prioritization;
●   Methods must be developed to retard degradation rates of operating pipelines, in order to forecast rehabilitation resource requirements.

Focus:
Research areas should focus on all aspects of pipeline degradation including: both oxidation and

microbiollogically induced corrosion (MIC), erosion, abrasion, hydrogen embrittlement, fatigue, and exterior force damage. The needed research should cover the entire range of pipeline applications, from domestic sewer and waterlines, to liquid petroleum product, gas transmission, and offshore facility related pipelines.

Scope:
- Field evaluation of in place pipelines - determine the variables;
- Laboratory load cell and in-situ tests of buried pipe in various configurations of degradation, such as:

    a. radial segment of wall thickness missing.
    b. inner reinforcing steel missing or partial steel missing in reinforced concrete pipe.
    c. invert abraded/thin wall in invert area.
    d. different types of bedding and backfill, cohesive vs granular soils.

Potential Funding Sources:
Potential funding sources include the DOT, public and private utilities and pipeline operators, pipe suppliers and manufacturers, NSF, EPA, TRB, AWWARF, GRI.

2.   LONG TERM PERFORMANCE OF PLASTIC PIPE

Define Long-term Characteristics:
Independent design limitations incorporating a broad range of actual insitu conditions for plastic conduits and their use is generally lacking in the industry. The general extrapolation of a standard 10,000 hour material test may not be adequate to properly ensure initial design of typical twenty-five to fifty years systems when specific conduit use and variable placement criteria has not been researched.

The long term ability of plastic pipe to resist buckling is a function of time, the condition and material properties of the host pipe, AND the specified plastic liner. Annulus grouting when incorporated in rehabilitation with plastic pipes adds yet another dimension needing thorough documentation and consideration by the design engineer so as to fully capture define or identify the most cost effective system and material requirements long term.

Only through additional new research incorporating the contribution of the host pipe existing conditions with that of the new plastic lining material can standardization of the evaluation in variable environments be quantified long term allowing for best selection of state of the art trenchless plastic pipe reconstruction systems and pipe specification.

Substantial Existing/Growing Investment
Over the past fifteen years, many plastic trenchless technologies have been developed and are being incorporated in our pipelines throughout the world. This new major plastic trenchless

pipeline reconstruction industry represents a major & growing investment. However, this new investment pales in comparison to initial new construction investment and its replacement cost.

Our national aging pipeline infrastructure assets must clearly be protected from technological standards advanced by specific self interest groups promoting a specific product. Independent plastic pipe material research that transcends all technologies is absolutely in our national self interest and is becoming a critical need for protection of this aging substantial insitu investment.

Relative Short History of Use:
Plastic pipe system rehabilitation has a relative short history of use when compared to other system material use. This relative short period of use coupled with insufficient research on plastic raw materials, compounds critical needs being cost effectively addressed with plastic pipe. Only through independent research can a wider acceptable comfort level be achieved for the use of plastics. Inherent characteristics of plastics such as high expansion and contraction need further research and documentation quantifying long term performance within a system; this will translate into acceptable known risks for the engineer and the utility owner along with delivering cost effective rehabilitation.

New Materials Being Developed:
Composite plastic materials are continually being developed. Research identifying desirable long term plastic performance can accelerate plastic pipe innovation resulting in cost effective rehabilitation of our assets.

Potential New Applications:
Potential new applications for plastic use will accelerate provided research documents critical long term characteristics. Gas, water, and power, along with sewerage systems have specific unique requirements. However, collectively all are concerned with long term performance of plastics pipes used in their systems. When more research is accomplished the application for new uses increases making even more effective research dollar spending.

Potential Funding:
DOE, GRI, NSF

3.   DETERMINING THE INTEGRITY OF CAST IRON PIPE

Structural Assessment Techniques for Cast Iron Pipe:
In general, a major share of water, sewer and gas utility investment is committed to transmission, collection and distribution pipelines. Water and gas pipelines in many cities in North America are old and consist primarily of unlined cast iron pipes. The lack of proper and timely maintenance of water distribution network systems is causing the incidence of pipe failures to increase at a high rate in many cities. Aging iron pipelines and consequent pipeline failures are a concern of utilities. A key factor in rehabilitating aging iron pipeline systems is the ability to assess the condition of buried pipelines. Only after the condition of the system is known can an evaluation of feasible rehabilitation measures be made. Water utilities in North America do not have technologies to directly assess structural conditions such as extent of graphitization, or wall thickness of a water pipeline.

When a water pipeline is installed, it is subject to structural stress due to loading from over-burden, frost, traffic, and other incidental loadings. The same pipeline also remains in contact with the surrounding soil environment outside and a water environment inside. These inside and outside environmental, physical, and chemical interactions may deteriorate the structural strength of the pipeline over a long period of time. These phenomena are complex and are not well understood. Development of a predictive model incorporating all possible environmental and outside loading interactions with a pipeline, and their long-term impact on structural condition of the pipeline, will help water utilities to understand long-term behavior of the pipeline and to maintain pipelines proactively.

Water utilities using surface water as source are experiencing a large number of breaks during a short period of time when temperature of water changes rapidly. Without any knowledge of phenomena causing these breaks, water utilities respond reactively to repair these breaks, resulting in severe disruptions of water service and large expenses for repair costs.

Condition Assessment Techniques:
There are a number of non-destructive techniques available for the assessment of the structural condition of a pipeline. Some of these techniques, such as intelligent pigging, have been used in the oil and gas industry. However, the water industry, with its special conditions, such as tuberculation of pipeline and water quality concerns, has been hesitant to try such techniques. It is necessary to study these issues and overcome barriers and adapt these available technologies. The needed research will identify specific concerns and barriers preventing the utilization of these technologies in the water industry; will develop solutions to overcome the barriers and concerns; will modify the existing technologies as necessary to make these technologies cost-effective to meet the needs of the water industry; and will demonstrate the applicability of these technologies by conducting several pilot projects.

Focus:

This research project will study all phenomena which interact with a buried iron pipeline with its outside and inside environments (such as soil corrosivity, bedding condition temperature, traffic and other loads, water characteristics, water temperature corrosivity, etc.); will incorporate these phenomena to formulate a predictive "pipeline condition assessment" model that describes the deterioration process of a water pipeline. The model, once developed, shall be verified by applying the model in actual case studies to predict deterioration of existing pipelines. Applicability of the model under various environmental conditions will be tested and environmental interaction parameters will be assessed. Induced stresses resulting from sudden change of water temperature causing main breaks shall be analyzed using the model and proactive maintenance remedies shall be identified.

Justification:
In order to rehabilitate and renew water system pipelines of North America, it is estimated that approximately $100 billion is needed over the next two decades. Development of structural assessment techniques applicable to water pipelines and formulation of a predictive water distribution pipeline condition assessment model describing deterioration process will help in:

- cost-effective allocation of resources and developing cost-effective rehabilitation programs saving millions of dollars, reducing interruption of services and reducing loss of water.
- prevention of catastrophic failure by proactive maintenance using the tools developed in the research
- properly maintaining and protecting existing investment of billions of dollars

Sources of Funding:
- National Science Foundation,
- American Water Works Research Foundation,
- DIPRA

4. STRUCTURAL REHABILITATION TECHNIQUES FOR EXISTING UTILITY PIPELINES

Introduction:
There is an aging infrastructure of gas, and water distribution pipelines in the older parts of many major chutes throughout the United States. Included in this aging infrastructure are about 55,000 miles of gas lines, and some 850,000 miles of water mains. These pipes are deteriorating at various rates due to a number of factors such as corrosion, adversely affecting their structural integrity, and resulting in failures. The cost of replacing these aging pipelines would be enormous. In addition, some areas have become so congested that it is virtually impossible to replace a buried pipe.

Justification:
Cement mortar lining has been the major non-structural rehabilitation method for water mains.

Structural rehabilitation has been limited to the insertion of steel liners in large diameter pipes. High Density Polyethylene Liners have been used extensively in Japan and Europe for structural rehabilitation of small diameter pipes. A few water utilities have implemented this technology in the United States. A number of factors, such as lack of adequate knowledge of long term physical and chemical performance of such liners, the impact of temperature changes on their behavior, as well as potential health hazards, have been the major impediments to their widespread use in this country.

The oil and gas industry, worldwide, has gained more experience with several structural rehabilitation techniques. For example, many gas lines have been rehabilitated with polyethylene liners. It is quite feasible to adapt these techniques for rehabilition of water mains.

The recent advances in the development of composite materials opens a new window of opportunity to develop new materials and rehabilitation techniques of aging pipes for the gas, oil, and water industries. These composite materials potentially offer flexibility, durability, high strength, corrosion resistance and impermeability.

Structural rehabilitation, using trenchless technology methods, can lead to extended useful life of buried pipes and minimize the failures. It is estimated that some $250 billion is needed over the next two decades to rehabilitate the aging water distribution and sewage pipelines of this country alone. There is therefore a genuine and urgent need to develop cost-effective rehabilitation techniques, if we are to solve this enormous problem.

Focus:
Three major research topics which are closely related to the structural rehabilitation of aging pipes are identified here:

1. An evaluation of existing rehabilitation methods in the gas, oil, and water industries.
2. The adaptation of currently used technologies in the gas and oil industry for use by water industry.
3. The development of new rehabilitation techniques for gas, oil, and water industries.

Scope:
Any research in the three topics mentioned here should include the following items:

1 A technical evaluation of the rehabilitation method
2 Increase in useful life of rehabilitated pipe
3 Design criteria and design methodology
4 Construction limitations/constraints
5 Specifications
6 Quality control and testing
7 Long-term performance of the materials involved
8 Health hazard potential of materials used for water mains

9   Compliance with applicable regulations
10   Costs
11   Environmental impacts

Potential Funding:
1 Department of Transportation
2 EPA
3 DOT/OPS
4 AWWARF
5 CERF
6 NSF

F.   Automatic Control and Instrumentation
Facilitator: William A. Hunt, MSE-HKM Engineering
The automatic control and instrumentation group was headed by Bill Hunt, Former chairman of the task committee on Automation of Water Pipeline Suystems. The group was made up of

1.   INSTRUMENTATION

Description:
Improve the accuracy and reliability of instrumentation used on pipelines. In particular, improvements to ultrasonic flow meters and on-line viscometers could directly or indirectly offer significant reduction in operating costs.

Justification:
The installation and maintenance of flow measurement devices and stations capable of producing the minimum expected 0.25 percent accuracy for transfer of custody of products is expensive and labor intensive. Likewise, current methods for measurement of on-line viscosity in petroleum lines is difficult and labor intensive as the viscometers must be removed and cleaned on the average of once per week. Non-intrusive accurate instruments for these functions are needed for improved efficiency of the industry. Improvements in this technology will make measurements simpler, less expensive to install and less prone to leakage. The non-intrusive instruments that exist, ultrasonic flowmeters, for example, cannot yet be used for highly accurate flow measurement. For those companies doing extensive modeling, real-time measurement of viscosity is necessary, yet no rugged non-intrusive instrument exists. Such an instrument will reduce the costs and due to inaccuracies of spot sampling of current practice.

Ultrasonic flowmeters are attractive as these are non-intrusive, not affected by build-up of biofilms and precipitates, have bi-directional flow capabilities, and are unaffected by pulsating flows.

Focus:
To examine the physical limitations and operating characteristics of the current devices to determine the feasibility of removing or overcoming these limitations by new approaches or technologies. The current state-of-the-art ultrasonic flow meters require extensive beta testing, establishing standards for testing procedures, development of standards of accuracy, analysis of electromagnetic interference on signals and telecommunication links to central data processing, development of methods for use with programmable logic controllers (PLC) and distributed control systems (DCS).

Funding:
Industry and vendors

2.   TRAINING SIMULATORS

Description:
Develop trainer simulation applications using software coupled to mock-up control panels. Research is needed to develop training methods and methods for auditing the effectiveness of simulator-based training.

Justification:
Industry and regulatory agencies have expressed considerable interest in pipeline simulators for training and updating operators on normal and emergency conditions. However, there are no clear criteria regarding the capabilities of and how well the actual pipeline operation is to be simulated. A poor training tool may actually diminish the performance of the trainees, as in the existing "craft" system used by many companies. Programmed scenarios and formalized training aids on the approach to using simulation training should be provided. In the end it may be possible to develop a performance standard for controller training, possibly a consensus standard could be developed in cooperation with the DOT-OPS. The trainer should b justifiable as a cost saving to the company. The goal of the industry should be towards having controllers who are so well trained that the risk of human error approaches zero.

Focus:
It is expected that the trainer simulators would be used to:
● test new procedures and familiarize pipeline controllers with new operations prior to implementation of new operating procedures or control strategies;
● qualify new pipeline controller personnel and requalify existing operating personnel;
● provide pipeline controllers with an opportunity to encounter all possible normal operating scenarios which may occur but which may be reported only infrequently;
● provide pipeline controllers and opportunity to learn to handle upset conditions and initiate emergency response procedures so thoroughly that the controller would develop a conditioned response to emergency situations.

Part of the research will involve technology transfer of training used in other industries and programs, i.e., NASA, the nuclear industry and airline pilot training.

Funding:
Joint industry/DOT-OPS funded activity with support from pipeline groups such as AWWA, API, GRI.

3. APPLICATION OF EXPERT SYSTEMS

Description:
Develop knowledge-based expert systems for pipeline operations utilizing commercially available expert system software. Encode available pipeline operating expertise to provide a tool to assist with decision making.

Justification:
Expert systems extend employee's capabilities by providing some of the expertise of others. Expert systems can be used, for example, to review large volumes of operating data, enabling the operators and managers to make decisions on alternatives for improving operational efficiency and/or proposed capital improvements based on historical information.

As most pipelines have operated for a number of years, a great deal of expertise resides within each company. For example, because of shift work in the direct operation of the pipelines handled through a control center, different pipeline controllers operate the line which leads to variations in operations, some of which may not be optimal. In this case an expert system should provide improved operations through offering:

- alternate choices which may be unfamiliar or uncommon to a controller;
- ability to utilized more data and to evaluate data more quickly;
- ability to monitor performance functions consistently without variance (such as caused by inattentiveness or fatigue on the part of the operator).

Another possible application of expert systems may be on-line power optimization. Maximizing the thoughput while minimizing the energy consumption is a very complex engineering process which could render great cost savings through the use of an expert system.

A third application the calibration of on-line instruments by using the rapid evaluation of SCADA system data rather than the current labor intensive methods.

Focus:
To start applying expert system methodologies to pipeline operations with the objective of improving operating efficiencies. It is suggested that initial applications be on small solvable problems that can create success stories and demonstrating a reduction in operating costs before embarking on grandiose scenarios.

Funding:
Primarily from industry. Expert systems applied to safety issues and risk management may be partially funded by DOT-OPS.

G: Freight Pipelines
      Facilitator: Henry Liu, Director, Capsule Pipeline Research Center

Background Information:
Recent advancements in pipeline technology have made it feasible to transport freight (solids) of various types in increasing volume and distance. The use of such freight pipelines not only enhances economic development and increases U.S. competitiveness in the world market, it also has far-reaching environmental and safety benefits to the nation including the following:

● Conservation of energy resources especially oil as a transportation fuel;
● Reduction in the number of trucks on the highway, hereby improving highway safety, reducing highway congestion and accidents, increasing the life of highway infrastructure and reducing highway maintenance cost;
● Reduction in the number of freight trains, thereby alleviating traffic jams and accidents at rail crossings, and saving lives;
● Decreasing air pollution and noise generated by trucks and freight trains;
● Transportation of freight underground, thereby improving landscape and land-use efficiency;
● Underground pipelines are weatherproof and most reliable. Theft of cargo during transport is also eliminated.

Due to the foregoing, increased use of pipeline for freight transportation is of national interest and should be a part of the overall government transportation and economic development policy of the future. Necessary research and development (R&D) to bring about increased use of freight pipelines should be encouraged by government agencies, especially the U.S. Department of Transportation, the Federal Highway Administration, the U.S. Department of Energy, the U.S. Department of Energy, the U.S. Department of Interior, and the U.S. Environmental Protection Agency. Five types of freight pipelines deserve attention in research:

(1)   tube transport,
(2)   coal log pipeline,
(3)   hydraulic capsule pipeline,
(4)   pneumatic pipelines, and,
(5)   slurry transport.

Each of the five are separately discussed and justified as follows:

1.   TUBE TRANSPORT
Description:
Tube transport is usually referred to in scientific literature as "pneumatic capsule (PCP)." It uses underground pipelines or tunnels to transport freight in cargo-carrying vehicles called "capsules." Any cargo of a size smaller than the capsule can be transported by such a system. Using pipelines of 1m diameter, many freights such as mail, food, and packages can be transported. Large systems of tube transport, using pipelines of 2m diameter of larger, can transport 90% of freight. Even larger tubes, in circular or rectangular cross sections, can be built and used to transport containers and trucks. However, sufficient freight demand is needed before such large tube

systems can be economical.

Research Needs:
Research needs in tube transport include designing, construction and testing a prototype system, testing proposed advanced propulsion systems, study of internodal transfer control technology, design of terminal facilities, economic and market analyses, and investigation of barriers to implementation.

Potential Funding Sources:
U.S. Department of Transportation, Federal Highway Administration. (Due to the high potential benefits of tube transport to highway systems, and the high cost and high risk in such research and development, it is only befitting for the federal government to undertake or sponsor such research. The missions of DOT and FHA are closest to that of such a new inter-city and interstate transportation system. The transport systems could be planned in conjunction with future highway projects.) This type of R&D is also suitable for the aerospace industry and defense laboratories interested in defense convention.

2.   COAL LOG PIPELINES

Description:
Coal log pipelines (CLP) is the transport of coal in cylindrical forms (called "coal logs) through underground pipelines from coal mines to power plants. In the United States, 60% of the electricity is generated from coal. The nation mines over one billion tons of coal each year, most of which must be transported for distances over 50 miles, sometimes over 1,000 miles. Due to the passage of the 1990 Clean Air Act amendment, most electric utilities in the Midwest such as Missouri have switched to the low-sulfur coal produced in Western states such as Wyoming. The coal sold at the mine in Wyoming costs less than $5 per ton. When transported to Missouri, it cost approximately $20 per ton. The difference is transportation cost. For a typical large power plant that uses 4 million tons of coal a year, a saving of $5 per ton in transportation cost saves the power plant 20 million dollars a year. Coal log pipeline can cause such large savings to power plants, thereby reducing the cost of generating electricity significantly. The public benefits from both reduced electricity cost and an improved environment. The latter stems from reduced use of coal trains and coal trucks. Five years of extensive R&D in coal log pipeline, sponsored by the National Science Foundation, U.S. Department of Energy, Missouri Department of Economic Development, and a consortium of 15 private companies, has brought the coal log pipeline technology close to commercialization. However, continued R&D for another three years, focused on large prototype tests, is needed before the technology is sufficiently reliable for commercial use.

Research Needs:
Remaining research needs in CLP are focused on the testing of a large coal log manufacturing machine and testing of coal logs produced by the machine in a large pipeline to allow accurate assessment of wear of coal logs in pipe. A pilot plant of an entire coal log pipeline facility is also needed to advance the CLP technology and accelerate its commercial use.

Potential Funding Sources:
The same government/industry consortium which is currently supporting the development of CLP. This includes the National Science Foundation (State/Industry University Research Program), Missouri Department of Economic Development, U.S. Department of Energy (Pittsburgh Energy Technology Center), and a consortium of 15 private companies. For the planned large-scale tests and pilot plant facility, additional funding from DOE and EPRI (Electric Power Research Institute) is needed.

3.   HYDRAULIC CAPSULE PIPELINES (HCP)

The transport of freight, such as grain or solid waste, contained within capsules (containers) that are suspended and propelled by water in a pipeline is a form of HCP. The coal log pipeline (CLP) is one type of HCP. Once the coal log pipeline development, currently in progress, is completed, the same technology can be adapted for transporting freight other than coal by using HCP. The same environmental and safety benefits of the tube transport and CLP pertain to HCP.

Research Needs:
The main research needs in HCP include propulsion systems (special booster pumps), hydraulics and hydrodynamics, drag reduction, and the investigation of neutrally buoyant capsules.

Potential Funding Sources:
Study of the use of HCP for transporting grain may be sponsored by large companies, cooperatives, and grain shippers and the U.S. Department of Agriculture. Studying the hydraulic, hydrodynamics and drag reduction of HCP may be sponsored by the Hydraulics and Particulates Program of the National Science Foundation, and investigation of the propulsion system of HCP and neutrally buoyant capsules may be sponsored by the Civil/Structure Program of NSF.

4   PNEUMATIC PIPELINES

Description:
Commonly referred to as "pneumatic conveying," pneumatic pipelines are in use extensively throughout the world for transporting hundreds of products including grain, coal, cement, plastic pellets, sand, chicken and hundreds of other products, such pipelines are relatively short, rarely more than a mile long.

Research Needs:
This is a mature technology that requires future research only in a few specific areas, such as new instrumentation and new methodology for measuring mass flow and concentration of solids, diagnosis of instability, system optimization, electrostatics and operational safety, and pipe erosion mitigation.

Potential Funding Sources:
General research in instrumentation, diagnosis of instability, electrostatics phenomenon, and pipe erosion study may be funded by National Science Foundation and the U.S. Department of Energy. Product specific research may be funded by specific industries. For instance, electric utilities and

EPRI may be interested in funding special studies on pneumatic conveying of coal powder or flash at power plants. Grain companies and U.S. Department of Agriculture may be interested in funding pneumatic conveying of grain.

5   SLURRY PIPELINE

Description:
A "slurry pipeline" transports solids in slurry or paste form. Slurry pipelines are used extensively throughout the world for transporting minerals and mineral wastes. They have also been used to a limited extent for transporting coal. A successful coal pipeline is the Black Mesa pipeline which transports 5 million tons of coal from Black Mesa, Arizona to Laughlin, Nevada.

Research Needs:
Research needs in the field of slurry pipelines includes slurry transport of mineral slugs (logs) in pipelines, rheology and pipeline transport of coal-water fuel which contains 70% coal and 30% water, and research to remove barriers on coal pipelines such as lack of eminent domain rights and right to cross railroads. The coal-water fuel is used as a liquid fuel to substitute for oil burned at power plants.

Other Recommendations:
In addition to discussing and prioritizing research needs in the freight pipeline area, the Group also feels that there is a general lack of public awareness, even within the engineering community, on the capability and advantages of transporting freight (solid cargoes) by underground pipelines. Such lack of awareness hinders the development of freight pipelines. Therefore, actions are needed by leaders of the engineering community, especially from the ASCE (American Society of Civil Engineers) and CERF (Civil Engineering Research Foundation), to educate the public about this new technology. One of the many actions suggested is to prepare a video tape, less than 10 minutes long, to educate the public about freight pipelines.

APPENDIX A

Bibliography of Suggested Research Topics

Group A: Pipeline Safety and Protection

Tom Steinbauer, The Gas Research Institute
    Detection of Emergency Situations on Natural Gas Transmission Pipelines
Richard Bonds, Ductile Iron Pipe Research Association
    Potable Water Pipeline Gasketed Joints Susceptibility to Permeation/Degradation
    Thrust Restraint Design for Flexible Joint Pipelines
Cesar DeLeon, Office of Pipeline Safety, Department of Transportation
    Rehabilitation of Aging Gas Distribution Pipelines
    Road Casing Research
Jim C. P. Liou, Department of Civil Engineering, University of Idaho
    Pipeline Leak Detection
    Integrity Assessment of Pipelines by Non-Disruptive Means
Robert Eiber, Consultant
    Risk Management
    Pipeline Maintenance
    Third Party Damage
Maher Nessim, Centre for Engineering Research
    Pipeline Risk Analysis Methodologies
    Limit States Design of Pipelines

Group B: Pipeline Design

Harry E. Stewart, School of Civil and Environmental Engineering, Cornell University
    Real Time Condition Monitoring of Pipelines
    Remote Detection of Pipelines
    Reliability of Distribution Systems
Dale F. Reid, Exxon Production Research Company
    Rapid Pipe Joining
    Reliability-Based Pipeline
    Insulated and/or Heat Traced Flowlines
Raymond Sterling, Trenchless Technology
    Improved Detection of Existing Services and Obstructions
    Improved Prediction of the Lifetime of Pipeline Repairs
    Improvements in Design and Construction in Microtunneling Pipeline Installations
Wes McGehee, Pipeline Engineering Consultant
    Basing Maximum Allowable Operating Pressure on Hydrostatic Test rather then
        on Specified Minimum Yield Strength (SMYS)
    Increasing the MAOP of Pipe by Use of Wirewrap
    Risk Management
Henry E. Topf, Jr., Miller Pipeline Corporation
    Standardization of Design and Material Applications for Trenchless Pipeline
        Reconstruction by an Independent Agency
    Service Reconnection
    Standardization and Streamlining the Process of New Product Evaluation by

Industry and by an Independent Technical Agency

Group C: Pipeline Operations

Thomas Hoelscher, Technical Manager Field Operations, Division III
Improve Methods to Locate Underground Utilities Without Excavation
Develop Better Coating for Buried Steel Pipe and More Efficient Methods of Installation
Improve the Quality of Information Obtained from Smart Pigs
James R. Lehman, Trunkline Gas Company
Damage Mitigation Offshore: Offshore One Call System, Safe Mooring Areas, Offshore Public Awareness, Emergency Plans
Brian Webb, BKW, Inc.
Effective Use of Auger Anchors
Mike Rickman, City of Dallas Water Utilities Department
Water Main Replacement Criteria
Water Main Rehabilitation
Mel Kanninen, MFK Consulting Services
More Widely Applicable, Accurate and Less Intrusive Inspection Methodologies
Quantification of Pipeline Damage/Degradation Mechanisms
Transmission Pipeline Rehabilitation Procedures

Group D: Fluid Mechanics/Hydraulics of Pipelines

E. B. Wylie, Department of Civil and Environmental Engineering, University of Michigan
Valve Dynamics - particularly check valves
Transient Flow Induced by Thermal Events
Leak Detection in - oil product lines natural gas lines water distribution systems
Thomas Walski, Wilkes University
Water Quality Changes in Water Distribution Systems
Manual Practice on Trenchless Technology
Frost Protection for Water System Components
John Bomba, Kvaerner - R. J. Brown
Calibration of Prediction Techniques for Slug Flow and Slug Length
Steven J. Troch, Baltimore Gas and Electric Company
Ultrasonic Weld Inspection
New Concept for Locating Underground Plastic Pipe
Composite Materials

Group E: Construction and Rehabilitation of Pipelines

Kenneth Kienow, Kienow Associates, Inc
Structural Adequacy of MIC Corroded Concrete Sewers
Friction Factors for Rehabilitated Pipelines
External Hydrostatic Long Term Buckling Resistance of Pipeline Rehabilitation Liners
James R. Baker, President Baker Pipeline

Deep Water Development - Gulf of Mexico
Environmental Impacts and Government Regulations
Trenchless Technology
Kent A. Alms, St. Louis County Water Company
Water Main Replacement Techniques
PE, PVC or Other Plastic Carrier Pipes
Corrosion Mitigation for Existing Pipes
Arun K. Deb, Roy F. Weston, Inc.
Water Main Breaks Due to Water Temperature Change
Water Main Renewal/Rehabilitation Program
Daniel W. Cook, Cook Construction Company, Inc.
Evaluation of Long Term Physical Characteristics of HDPE & PVC Conditions
Development of Remote Pipeline Condition Assessment Equipment for Evaluating
Steel, Ductile Iron and Cast Iron Pipes
Develop a Trenchless Technology Public Information/Educational Committee
within ASCE to Promote the Advantages and Social Cost Savings to the
Public
Ahmad Habibian, American Society of Civil Engineers
Development of Condition Assessment Technologies
for Water Mains
Structural Rehabilitation of Cast Iron Water Mains
Earthquake Hazard Effects

Group F: Automatic Control and Instrumentation
William A. Hunt, MSE-HKM Engineering
Ultrasonic Meters for Gases and Liquids
Leak Detection Processing from SCADA Systems
Formulation of Program Logic Controllers (PLC) Responses to Fluid Transients
Aubrey F. Zey, NovaTech,
Leak Detection
Power Optimization
Electronic Data Interchange
Ed Farmer, EFA Technologies, Inc.
Emergency Response Plan Implementation
Automatic Acting and Remote - Controlled Line Block Valves
Valve of Training Simulators
F. Roy Fleet, Natural Gas Pipeline
Improvements to Line Break Detectors
William F. Quinn, El Paso Natural Gas Company
Automatic Control Valves (ACV) and Remotely Controlled Valves (RCV) for
Mainline Gas Transmission Pipelines
Excess Flow Valves (EFV) for Gas Distribution Service Lines
Remote Monitoring and Control of Pipeline Systems
Don Scott, Interprovincial Pipe Line Co.
Application of Expert Systems

Instrumentation

Group G: Freight Pipelines

Henry Liu, Capsule Pipeline Research Center
Coal Log Pipeline
Hydraulic Capsule Pipelines (HCP)
Pneumatic Capsule Pipe (PCP)
Recommended by: William Vandersteel, Ampower Corporation
Tube Freight
Recommended by: Lawrence Vance, US Department of Transportation
Economic Feasibility of Pneumatic Capsule Pipeline
Terminal Design
Market Analysis for Pneumatic Capsule Pipelines
Sean Plasynski, US Department of Energy
Pneumatic Transport (Dilute and Dense Phase Solids Transport)
Coal Log Pipeline
David T. Kao, Iowa State University
Energy Efficient Capsule Freight Pipeline Transport System Development
Hydraulic Capsule Pipeline Transport System as an Integral Component of Freight
Transport Network in Developing Regions
Development of Techniques for In Situ Pipeline Transport Infrastructure Failure
Detection and Rehabilitation
Thomas J. Pasko, Federal Highway Administration
Freight Movement in Tunnels

Group A: Pipeline Safety and Protection

Includes pigging for safety and integrity, pipeline leak detection and monitoring, pipeline spills, effects of earthquakes, hurricanes, and floods on pipeline safety, cathodic protection systems, and third party damage prevention.

Facilitator:

Tom Steinbauer
Gas Research Institute
8600 West Bryn Mawr Avenue
Chicago, IL 60631
(312) 399-8100

Panelists:

Terry Boss
Interstate Natural Gas Association of America
Suite 300 West 555 13th Street, NW
Washington, DC 20004
(202) 626-3234

Richard W. Bonds
Ductile Iron Pipe
Research Association
245 Riverchase Parkway
Birmingham, AL 35244
(205) 988-9870

Cesar DeLeon
US Department of Transportation
Office of Pipeline Safety Research
400 Seventh Street, SW
Washington, DC 20590
(202) 366-4595

Jim Liou
Department of Civil Engineering
University of Idaho
Moscow, ID 83843
(208) 885-6782

Bud Danenberger
MMS
Engineering and Technology Division
381 Elden Street, Mail Stop 4700
Herndon, VA 22070-4817
(703) 787-1559

Robert Eiber
Pipeline Consultant
4062 Fairfax Drive
Columbus, OH 43220
(216) 538-0347

Maher Nessim
Centre for Engineering Research
200 Carl Clark Road
Edmonton, Alberta
Canada T6N 1E2
(403) 450-3300

Recommended by: Tom Steinbauer, The Gas Research Institute

Detection of Emergency Situations on Natural Gas Transmission Pipelines

One of the lingering concerns of the public affected by the March 24, 994 pipeline failure near Edison, New Jersey is the inability to automatically close isolation valves in the event of a natural gas pipeline failure. The implication is that the immediate closing of isolation valves adjacent to a pipeline rupture would have prevented at least a portion of the devastation realized in Edison New Jersey on that fateful day.

There are many issues that one must consider to properly focus this concern. A few of these are:

• The vast majority of damage that results from a gas pipeline rupture or explosion occurs within seconds given the compressible nature of the product being transported. Historically there has been very little additional damage caused by the escaping product subsequent to the initial rupture. Consequently, the main beneficiary of the rapid closing of isolation valves is the pipeline company, as closing these valves prevents the escape of valuable product.

• There is no technology currently available that can reliably sense and signal valves to operate in the event of a gas pipeline rupture. As a July 1995 report by the Gas Research Institute (GRI-95-0101) points out, current line-break technology has an unacceptably low 50% success rate if we are to believe the records of one of the major US gas pipeline companies that has state-of-the-art line break control technology on virtually all of its isolation valves. Thus new technologies must be developed before automatic and/or remote operation isolation valves can be made reliable.

• A reliable means for detecting leaks and/or ruptures (which, in the minds of many are one and the same thing) is simply unknown to the gas pipeline industry at this time. Thus even if reliably operating valves and peripherals were available, reliability would still be sacrificed given the lack of any indication that pipeline failure had occurred, and the magnitude of that failure (leak vs. rupture).

• Historically speaking, the vast majority of pipeline failures have resulted from third party damage; as was the case with the New Jersey incident. Therefore the gas pipeline industry needs to know not only when third parties damage their facilities, but also when third parties even encroach on their right-of-way.

An analysis of all of the factors surrounding these aspects of pipeline failures leads to at least the following conclusions:
• First of all, basic research is required to develop the technology to accurately and reliably detect pipeline leaks and ruptures so that isolation valves can be operated in the event of failure. Secondly, a reliable practical means for real time detection of third

party encroachments must also be developed. Only then can the concerns of the Edison New Jersey inhabitants begin to be addressed.

Recommended by: Richard Bonds, Ductile Iron Pipe Research Association

Potable Water Pipeline Gasketed Joints Susceptibility to Permeation/Degradation

Justification: Permeation of water distribution systems is a subject of continuing concern and investigation by water utilities, the American Water Works Association and others. Many types of pipe and gasket materials are used in water systems. The selection of materials is critical for water service and distribution piping in locations where there is likelihood the pipe will be exposed to significant concentrations of pollutants such as low-molecular weight petroleum products or organic solvents and their vapors.

Research has documented that pipe materials such as polyethylene, polybutylene, polyvinyl chloride and asbestos cement may be subject to permeation by low-molecular weight organic solvents or petroleum products; however, research regarding gasketed joints is limited. Tests have been conducted on thin films cut from gasket materials; however , this is not representative of actual gasketed joints (mass, compression, etc.). Needed research includes permeation/degradation tests on potable water gasketed joints for varying concentrations.

Thrust Restraint Design for Flexible Joint Pipelines

Justification: Restrained joints are now specified and/or installed where it is either impractical or undesirable to utilize concrete reaction blocks, or where additional joint security against joint separation or over-deflection is desired. To date, most of the data used to develop the current design approach has been based on tests of small diameter pipe (12-inch diameter and smaller). Conducting field tests on large diameter pipe would be both time consuming and expensive. Needed research would include finite element analysis of thrust restraint for flexible joint pipelines in order to determine the current used equation's conservatism, predict movement of piping system, effect of pipe diameter, etc.

83

Recommended by: Cesar DeLeon, Office of Pipeline Safety, Department of Transportation

Rehabilitation of Aging Gas Distribution Pipelines

Justification: There is an aging infrastructure of gas distribution pipelines in the older sections of many major cities throughout the country. Many of these pipelines are approaching or are already 100 years old. Included in this aging infrastructure are about 55,000 miles of cast iron distribution mains and some cast iron gas distribution service lines going to houses. Many of these cast iron pipes are susceptible to the corrosion process of graphitization and unanticipated brittle failure that requires constant monitoring. The costs to replace these aging pipelines would be extraordinarily expensive.

There is a need for a broad study regarding the efficacy of rehabilitation methods of distribution mains and service pipelines, particularly cast iron pipelines. The study would report on the technical merits of the various methods of in-situ replacement of metal pipe in poor condition with polyethylene plastic pipe or plastic liners which bond to the inside surface of the aging metal pipe. The study would establish design criteria for plastic pipe liners; determine if these liners constitute composite pipe (metal jacket and plastic liner) for design purposes or if the plastic liner can be designed to contain the internal pressure. Predict the operational life of the plastic liners. Determine if these rehabilitation methods comply with Federal pipeline safety regulations. Research the state-of-the-art and safety considerations of trenchless rehabilitation methods.

Road Casing Research

Justification: Most gas transmission and petroleum pipelines were constructed in casings when those pipelines cross highways and railroads. Many pipelines fail in the casings due to external corrosion as moisture accumulates in the annulus between the casing and the gas or petroleum pipeline. This occurred in Beaumont, KY in 1985 resulting in 5 fatalities. There is need for a study to determine if shorted casings (pipeline contacts the casing and shorts the cathodic protection current), shielding of the cathodic protection applied to the gas or petroleum pipeline, or casing annular space not filled with high dielectric material detrimentally affects the pipeline and, if so, determine technological methods to reduce risk of external, atmospheric, and chemically-induced corrosion.

Recommended by: Jim C. P. Liou, Department of Civil Engineering, University of Idaho

Pipeline Leak Detection

Justification: Because pipelines are buried and out of sight, they are subject to accidental third party damage. Amongst the main threats to the safety of oil and gas pipelines, third party damage poses the greatest risk in terms of the likelihood of occurrence and the consequences of failure. Software-based real-time leak detection can reduce this risk at a reasonable cost. This technology is still evolving and there is a wide range of sophistication. Although the effectiveness of leak detection schemes has been demonstrated for some pipelines, many leak detection systems fail to meet expectations. Difficulties in modeling, locating a leak, dealing with uncertain and noisy data, reducing false alarm occurrences, and implementation still remain. The potential of this technology is yet to be fully developed and explored.

Future research will establish the realistic capabilities and limitations of software-based real-time leak detection. The results will (1) help the pipeline industry to use appropriate methodology to reduce its safety risk from third party damage, (2) provide a rational basis for rule making by regulatory agencies.

Integrity Assessment of Pipelines by Non-Disruptive Means

Justification: Transmission mains are subject to hydrostatic testing before service to avoid failure due to defects. However, defects that survived the initial tests may grow and new defects may develop during service. Thus, the integrity of pipeline systems needs to be assessed periodically. This need is urgent for our aging pipeline infrastructure. Integrity assessment is being done by means ranging from aerial surveys (low cost, non-disruptive, but of limited value) to hydrostatic testing (high cost, disruptive, but comprehensive). Among these, intelligent pigs emerge as the method of choice for some oil and gas pipelines as it can be used on operating pipelines. However, intelligent pigs are expensive to run (several thousand dollars per mile), and the results may be unreliable. Furthermore, forty-two percent of natural gas pipelines, eleven percent of oil pipelines, and all water mains in the US cannot handle pigs due to physical limitations. In terms of cost and capability, there is a gap in the available tools for pipeline integrity assessment.

A preliminary study suggests that noise level pseudo-random pressure signals can be used to extract on operating pipeline's impulse response function. This function can then be used to assess the integrity of the pipeline in real time prior to failure. The technology will enable on-line real time monitoring of pipeline integrity in ways similar to monitoring operating machines. Besides its own utility, this technology will also help to decide when to do pig inspection and what type pigs to use. This technology may fill the gap stated above and contribute to the safe operation of aging pipelines.

85

Recommended by: Robert Eiber, Consultant

Risk Management

Justification: An area where research is just starting in the US is risk management on high pressure gas and hazardous liquid pipelines to determine means of controlling the exposure to the public and still provide the energy needed for homes and industry on an economical basis. The justification comes partially from the Edison, NJ failure in which the public was exposed to a potentially serious incident that fortunately did not result in catastrophic consequences. The same situation also probably holds true for other types of pipelines such as water mains. All can pose a risk to the public either through rupture, fire, environmental contamination, sink holes, flooding, etc.

If technology was sufficiently advanced that we could rule out the possibility of an incident on a pipeline then we would not need risk management. However, as one who has been developing technology for improved pipeline integrity, this is a slow process and the environment demands protection for the public in the near term. I believe the most immediate improvement can come from the application of risk management to pipelines. One problem with this statement is that risk management concepts and application are also in their infancy with regard to this application. Fortunately, other industries have been using the tools and therefore the concepts are available.

The natural gas industry is working with The Gas Research Institute on a risk assessment program and INGAA is working with DOT and API to explore and hopefully develop assessment methodologies and approaches. After reviewing one of the early GRI reports on risk terminology, I am hesitant to use a term like "risk management", which is the overall process of risk analysis and risk evaluation leading to risk management, because the language can be misinterpreted in so many ways. This is a start but it needs major attention and focus. There will be no one single approach that will work for all companies and many companies will need assistance, especially the smaller companies with limited resources and few pipelines. The overall goal should be to reduce the level of exposure to the public to a reasonable level. In todays' world there is always some risk associated with living in confined areas but that risk should not be at an unreasonable level.

It seems to me the goal of the breakout session in this area should be to determine:
1. What should be done to bring companies on board with the concept?
2. What should be done to educate companies on the methodologies/concepts that are available now and that can be applied?
3. How to get started.
4. What is needed to fully develop this technology for pipelines? What data needs to be collected? Who should collect?

Pipeline Maintenance

Justification: The aging infrastructure in US pipelines needs attention, just as it has in other industries. In the year 2000, over 50 percent of the high pressure gas pipelines will be over 40 years old. Age, of and by itself, does not cause degradation of a steel pipeline, but the effects of the environment can cause degradation of lines. In this area, it seems that the goal of the breakout session might be to define what maintenance activities should be addressed in a national research program. Topics that might be included are:

- Inspection techniques including In-Line Inspection tools to locate all of the defects that cause serious service incidents
- Integrity methodologies to assess defects/imperfections that are located,
- Run/repair/replace strategies for the industry so that minimum standards can be established for the future operation of pipelines.

Third Party Damage

Justification: Unreported damage to pipelines continues to be the major cause of incidents on both gas and liquid pipelines. With the increasing use of underground tunneling devices to lay small pipelines and with the increasing proximity of centers of population to pipelines, it seems to me there is a need to develop a means to detect when third-party damage is evident and prevent it's occurrence or at least detect that it has occurred.

The worst situation is when a pipeline is damaged and not reported. Depending on the severity of the damage, a pipeline can fail immediately or after years of continuing service during which time the damage slowly increases in severity until it fails when the public may be in close proximity. Damage that fails immediately, poses a threat to the people producing the damage and has caused a number of injuries and fatalities.

In this area, I could envision that the first goal of the breakout session should be to determine if there is consensus on this area of research. I am familiar with the incident statistics on gas and liquid pipelines and from 20 to 40 percent of the incidents that occur are from third party damage. I suspect that this may also be true with water lines. The second goal might be to establish objectives for the program, i.e., What types of approaches might be reasonable, such as instrumenting all excavation equipment, placing utilities in a utility corridor encased in concrete, instrumenting the pipelines to detect damage, etc.

Recommended by: Maher Nessim, Centre for Engineering Research

Pipeline Risk Analysis Methodologies

Justification: Risk analysis is gaining increasing recognition in the pipeline industry as a rational methodology to make operational and maintenance decisions that meet safety goals at the lowest possible cost. Available risk analysis approaches can be categorized into two classes: risk index methods and quantitative risk analysis (QRA). Risk index methods rely on subjective judgement and are often used to provide preliminary risk rankings of pipeline segments. QRA methods provide more objective risk estimates that can be compared directly to tolerable risk levels and used as a means of making optimal operational decisions. Existing QRA approaches use historical failure rates as direct estimates of the failure probability. Because of the rarity of pipeline, this approach gives generic failure probability estimates that cannot account for the specific attributes for a given pipeline or quantify the risk reduction associated with potential design and maintenance choices.

To resolve these problems, a model-based approach to QRA is required. This approach derives the probability of failure from pipeline condition data (e.g., the number and severity of corrosion features) and mechanical models that describe pipe failure conditions (e.g., the failure pressure of a pipeline with a corrosion feature of a given size). C-FER has an ongoing research program sponsored by a number of pipeline companies and regulatory agencies to develop the required models and facilitate their use in decision making. More research is required in this area over the next few years to 1) develop models required to implement model-based QRA for different pipeline failure causes including corrosion, mechanical damage, dent/gouges, cracks, and ground movements; 2) collect the pipeline condition data required for these models; and 3) develop tools that reduce the level of effort and specialized knowledge required to use risk analysis in every day decision making.

Limit State Design of Pipelines

Justification: Over the past two decades, many structural design codes have been converted into the reliability-based limit states format. This has proven to be a rational design approach that ensures adequate and consistent safety level against the relevant failure modes. Although, pipeline design codes in North American are still based on working stress design methods, there is interest in moving to a limit states design format. Some work on this topic has been sponsored by such organizations as PRCI and the National Energy Board of Canada. This topic may be seen as a design issue, however, it is also a safety issue in as much as limit states design is the best approach to achieving adequate safety.

The work carried out in this area has been preliminary in nature, and a significant amount of research is still required to 1) develop a consistent design philosophy for pipelines (e.g., should pipelines be designed explicitly for mechanical impacts or for time dependent deterioration mechanisms such as corrosion?); 2) characterize the degree uncertainty associated with different loading types such as excavator impact and internal pressure; and 3) calibrate a

88

set of limit states design safety factors. Equally important is the creation of a mechanism that permits transfer of this technology to pipeline engineers and obtaining their feed back in order to ensure that the results are useful to and accepted by the industry.

Group B: Pipeline Design

Includes new design approaches for onshore and offshore pipelines (including earthquakes, flooding, etc.), possible revisions to design codes, and use of expert systems in design.

Facilitator:
Raymond Sterling
Trenchless Technology Center
Louisiana Tech University
P.O. Box 10348
Ruston, LA 71272
(318) 257-4072

Panelists:

Harry Stewart
School of Civil and
Environmental Engineering
Cornell University
Ithaca, NY 14853
(607) 255-4734

Dale Reid
Exxon Production
Research Company
P.O. Box 2189
Houston, TX 77252
(713) 966-6174

Ibrahim Konuk
National Energy Board
of Canada
5th Floor 311 - 6th Avenue
Calgary, Alberta
T2P 3H2
(403) 292-6911

Wesley B. McGehee
Pipeline Engineering Consultant
14405 Walter Road, Suite 351
Houston, TX 77014
(713) 893-3080

Henry Topf, Jr.
Miller Pipeline Corporation
P.O. Box 34141
Indianapolis, IN 46234

Charles Smith
MMS
Research Program Manager
381 Elden Street
Mail Stop 4700
Herndon, VA 22070

Recommended by: Harry E. Stewart, School of Civil and Environmental Engineering, Cornell University

Real Time Condition Monitoring of Pipelines

Justification: Real time condition monitoring of pipelines conveying potentially hazardous substances is critically important for safety, environmental protection, and cost-effective operations. Gas and petroleum industry statistics show that the principal cause of pipeline accidents is third party damage, with approximately 40% of all reportable accidents attributable to hits and under mining of pipelines due to construction activities. Advances in microsensor and signal processing technology provide an opportunity for detecting third party incursions in real time, and for relaying the location of potential damage to a central monitoring system

The objectives of a focused research program in this area would be to 1) investigate and quantify the use of microelectromechanical and fiber optical systems in detecting low amplitude acoustic signals, 2) evaluate the dynamic characteristics of third party incursion signals , including frequency spectra, attenuation, and influence of pipe properties, soil types, and pipeline configurations on signal transmission, 3) develop and demonstrate opto-electronic capabilities for signal transmission and sensing and 4) design and demonstrate a prototype signal detection system.

Remote Detection of Pipelines

Justification: A significant portion of the utility network is located beneath paved or landscaped surfaces. Since the cost of reinstating these areas can be very high, trenchless installation methods often are preferred. The risk, whether perceived or actual, of hitting and damaging underground services or of encountering unexpected obstacles, constrains more widespread use of trenchless techniques. Overcoming this constraint requires the development of reliable methods for detecting underground conduits. Several geophysical approaches are available, but have practical limitations on the depths that can be probed, the inability to locate non-metallic objects, electrical interference from nearby fields, and difficulty in providing precise locations.

The objectives of this research would be to evaluate the potential of using several types of combinations of types of sensors to detect buried pipelines. Electromagnetic, magnetometer, seismic, acoustic, radar, and infrared sensor technologies would be evaluated. While commercial development of several of these sensor technologies has been pursued, there have been no definitive field confirmations of their suitability. In addition, there are new technologies being developed at several of the US National Research Laboratories that may show potential for use in detecting buried piping.

Reliability of Distribution Systems

Justification: Bare steel and cast iron systems must be evaluated for integrity, and procedures established for replacement and/or retention decisions. Often, decisions to replace or retain piping is made based on limited historic performance data. System-wide information on the nature and frequency of repairs, the suspected cause of the failure, and performance of the pipeline after being repaired is difficult to obtain. Thus, rational decisions as to the effectiveness of a repair strategy often cannot be made, and decisions are made to replace pipeline segments which may continue to provide reliable product delivery.

The objectives of this research would be to develop an approach to evaluating overall systems reliability using modern information management methods for data collection and retrieval, implementing GIS systems for complex and varied distribution networks, and using engineering techniques such as those common in the Operations Research and Industrial Engineering fields to evaluate replacement/retention strategies, the effectiveness of repair procedures, and how best to evaluate and improve overall system reliability.

Recommended by: Dale F. Reid, Exxon Production Research Company

Rapid Pipe Joining

Justification: Rapid pipe joining refers to new welding processes or mechanical joining methods that will permit faster offshore pipe joining compared to conventional welding. The offshore industry is continuing to move to deeper water that may require pipelines to be exclusively laid using a method called "Jlay". Using this method, pipe is stalked and welded in a near-vertical orientation from a construction vessel; all pipe joining, welding, inspection and coatings are applied at a single station. Therefore, productivity and pipelay costs are determined by the speed of these operations. If rapid welding processes or mechanical connectors could be made available with performance and reliability comparable to conventional welding, then 25-40% savings on installed costs may be realized. There is also incentive to use mechanically-joined pipe to permit the use of low-cost but non-weldable corrosion resistant alloys such as Cr13 ("Chrome-Thirteen") pipe. Such technology would be applicable to deep and shallow water applications that normally require higher cost CRA materials such as duplex stainless steel. If a mechanical connector technology is made available for CRA flowline applications, then cost savings on the order of 40-50% may be realized.

Reliability-Based Pipeline Design

Justification: Offshore pipelines are generally designed in accordance with established stress-based design codes. These codes have proven to be safe and conservative for conventional offshore pipelines, say for example, oil and gas transportation lines in water depths under 2000 ft. However, more and more offshore developments are requiring non-standard pipelines and flowlines, higher-strength materials, and design service and installation conditions that are quite different from what the codes were originally designed to address. Examples include subsea flowlines for high temperature, high-pressure service (300 F, 10,000 psi) and ultra deepwater pipelines (3,000-6,000 ft.). As a result, existing design codes may be inadequate to address these types of designs. There is incentive to develop new design procedures and codes based on limit-state and reliability-based design methods to replace or supplement traditional codes and design methods.

Insulated and/or Heat Traced Flowlines

Justification: Offshore developments based on oil and gas production from subsea wells often require well-insulated flowlines to prevent the occurrence of hydrate plugs and wax deposition. This is particularly the case with deepwater production. However, the high ambient hydrostatic pressures in deep water preclude the use of conventional flowline insulation technologies. New flowline insulation and heat tracing technologies are sought to aid the viability and economic development of deepwater oil and gas reserves.

Recommended by: Raymond Sterling, Trenchless Technology Center

Improved Detection of Existing Services and Obstructions

Justification: With the increase of installation of pipelines by trenchless methods, the ability to identify and locate existing services and obstructions is becoming more critical. Either damage to existing services or difficulty in completing planned installations can result from poor quality information in this regard. Improvement are needed in the technologies available in the site investigation phase and also during the drilling and installation phase itself.

Improved Prediction of the Lifetime of Pipeline Repairs

Justification: Many methods of trenchless pipeline repair and rehabilitation have emerged in the last decade or so. There is still, however, insufficient long-term experience with these methods to fully understand their long-term performance. The materials used for the repairs are often shaped and cured in the field and different products within the same general class of repair technique may use substantially different materials. Traditional materials testing techniques for acceptance and quality control purposes do not provide the necessary information on the performance of the field assemblies but accelerated testing of simulated field assemblies gives a wide variation in performance. Better performance models for design with appropriate safety factors and appropriate product testing requirements should be the result of the research.

Improvements in Design and Construction in Microtunneling Pipeline Installations

Justification: Pipelines installed by microtunneling or other pipe jacking methods are subject to radically different loading conditions than pipelines installed by traditional trench installation. The need to understand the nature of these loading conditions has been mitigated by the fact that the strength of the pipe has generally been controlled by its required thrust capacity to overcome the pipe skin friction during jacking. Improvements in lubrication muds used to reduce pipe friction are reducing thrust requirements and may allow thinner and less expensive pipe wall thicknesses to be used where the thickness is not also required for transverse loading resistance. Also, there have been a number of cases where microtunneling installations have run into great difficulties but there has not been an effective effort whereby the industry can learn from its mistakes and avoid them in the future. A continued effort to develop better pipe design procedures tailored to microtunneling and pipe jacking and to track installation and performance problems to pinpoint their causes would be valuable.

95

Recommended by: Wes McGehee, Pipeline Engineering Consultant

Basing Maximum Allowable Operating Pressure (MAOP) on Hydrostatic Test Rather than on Specified Minimum Yield Strength (SMYS).

Justification: A great amount of research has been done on basing the MAOP of pipelines on hydrostatic test rather than on SMYS. This research should be revisited and brought to the forefront for design criteria. There is no technical basis to limit MAOP of gas pipelines to 72 percent of SMYS in rural areas but there is voluminous data to support at least 80 percent SMYS based on hydrostatic test and other criteria. The ASME B31.8 has adopted this principal.

Increasing the MAOP of Pipe by Use of Wirewrap

Justification: This concept was developed many years ago but was aborted by Federal Regulations. Some research was done but was supported by documentation that may not be readily available. More research should be done to determine the viability of using wirewrap to increase the MAOP of pipe.

Risk Management

Justification: More research is needed to develop risk management program based on a quality assurance program in the design phase to enhance the pipeline system to one that will be safe for the public both initially and ongoing after installation. Also, research is needed to develop a monitoring system to alert the operator that a disturbance is occurring near the pipeline and an alarm to alert the third party excavator that they are near a pipeline. This technology should be incorporated in the design phase.

Recommended by: Henry E. Topf, Jr., Miller Pipeline Corporation

Standardization of Design and Material Applications for Trenchless Pipeline Reconstruction by an Independent Agency

Justification: Over the past fifteen years many trenchless technologies have been developed and are commercially installed throughout the world in a number of industries. Many of the standards for these technologies were advanced by specific self interest groups to promote a specific product.

The entire trenchless pipeline reconstruction technology has developed into a major industry including pipe replacement, pipe construction and pipe lining.

Independent research should be carried out that transcends the major industries and consolidates the technologies by defining the major rehabilitation categories, evaluating the stress that materials are subjected to during installations in each category and developing appropriate technically generated standards and equality of dissemination.

Service Reconnection

Justification: The reconnection of service pipes to a newly reconstructed mainline pipe will differ between industries. In gravity systems, a good leak proof seal is desired to minimize infiltration into the newly constructed system. In water pipelines, a leakproof seal is required to prevent water from being lost at the connection and in natural gas system, a leakproof connection will prevent gas migration between the newly installed pipeline and the host pipe preventing potential catastrophic occurrence.

Research is needed to technically evaluate all existing technologies available worldwide and to develop a standard of practice that will interface with the state of the art trenchless pipe reconstruction systems currently being installed in the various industries.

Standardization and Streamlining the Process of New Product Evaluation by Industry and by an Independent Technical Agency

Justification: New trenchless technologies are emerging weekly. Many are extremely promising but must be separately evaluated by specific local agencies around the country causing duplicity of effort and cost and lacking in scientific approach.

The evaluation process varies from location to location and in some cases may take several years before approval is obtained. Research is needed to standardize the evaluation process as it applies to a specific industry. The evaluation should be conducted based on technical criteria associated with design, construction and material properties after product installation, not by self interest parties having sufficient political influence.

97

Group C: Pipeline Operations

Includes any non-safety related operational issue such as use of drag-reducing additives to reduce power consumption, handling of emergencies and spills, economics of pipelines, pumping operation procedures, and maintenance of aging pipeline systems. Includes design for and use of pigs for pipeline cleaning, sizing, and entry ports, use of various instruments to detect pigs, leaks and corrosion. How to cope with hydrate formation and parafin build-up problems, particularly in deepwater oil and gas flowlines, will be explored.

Facilitator:

Tom Hoelscher
Transco
P.O. Box 7707
Charlottesville, VA 22906
(804) 973-4384

Panelists:

J. R. Lehman
Trunkline Gas Company
P.O. Box 1642
Houston, TX 77251
(318) 836-5689

Brian C. Webb
BKW, Inc.
P.O. Box 581611
Tulsa, OK 74158
(918) 584-4402

Mike Rickman
City of Dallas
City Hall 4A North
1500 Marilla
Dallas, TX 75201
(214) 670-8007

John Elwood
Foothills Pipe Lines Co.
3100 - 707 Eighth Avenue
Calgary, Alberta T2P 3W8
(403) 294-4137

Mel Kanninen
MFK Consulting Services
7322 Ashton Place
San Antonio, TX 78229
(210) 349-9882

Alex Alvarado
MMS
1201 Elmwood Park Boulevard
Mail Stop 5232
New Orleans, LA 70123
(504) 736-2547

Recommended by Thomas Hoelscher, Technical Manager Field Operations, Division III

Improve Methods to Locate Underground Utilities Without Excavation

Justification: Fifty percent of pipeline failures are a direct result of third party construction damage. A major component of the problem can be addressed by increasing awareness of buried utilities and public education about One Call Systems. Enforcement of the One Call law and punitive damages for those who do not use it will help. Too much time and money is spend by utility operators in locating their own facilities to ensure that new construction doesn't encroach on existing facilities.

The best pipe locating technology today uses an electrical signal put on steel pipe by a portable source through an above ground connection to the pipe. The receiving device picks up the signal from the pipe, up to a distance from the source of a mile, and indicates the location. The receiver can also give an indication of the depth of cover over the pipe. Unfortunately, excavation is still the only way to ensure that the horizontal and vertical location are known accurately enough to proceed comfortably with construction.

Develop Better Coating for Buried Steel Pipe and More Efficient Methods of Installation

Justification: Buried steel pipelines require coating to protect them from corrosion. When the coating fails, electrical currents are impressed from a remote location to reduce the amount of iron loss, helping to maintain the integrity of the pipe. The life span of some coatings, particularly the asphalt based coatings popular in the 50's and 60's, is limited. Eventually, the cost of cathodic protection becomes prohibitive, and the pipe must be recoated. This is a very time consuming, labor intensive project.

If tape is the coating of choice, we need to develop a mechanized system to remove failing coating and apply new and better coating with the pipe remaining in the ditch, in service. Other coatings require very specific conditions (temperature, humidity, cure time) that are difficult to find on pipeline ROW's for the length of time required.

Improve the Quality of Information Obtained from Smart Pigs

Justification: Current smart pig technology uses a magnetic field and passes it through a short section of pipe. The strength and consistency of the field is measured, with inconsistencies indicating anomalies in the pipe wall. Unfortunately, most of the anomalies identified are not defined well enough to clearly identify the problem, and the pipeline must be dug up and physically investigated to ensure the integrity at each separate location. With more definitive information, at a reasonable cost, many more miles of pipeline can be smart pigged in the future.

Recommended by: James R. Lehman, Trunkline Gas Company

Damage Mitigation Offshore: Offshore One Call System, Safe Mooring Areas, Offshore Public Awareness, Emergency Plans

Justifications: Reference the statistics in the Marine Board Safety Report. The justifications are to reduce the pipeline facilities accident rate, to reduce the environmental impact of offshore accidents, to reduce the fatalities due to such accidents, to reduce the fatalities and injuries to maritime users in the offshore environment.

The damage mitigation program offshore would consist of a multifaceted program to prevent and reduce the seriousness of accidents offshore. The program can best be accomplished through a partnership with the oil and gas pipelines, government, maritime industries, shipping, fishing, drilling, etc. The rewards are a win/win situation for the government and industries. Several facets of the program are discussed below.

Offshore One Call System:

Several attempts have been made to form a task group to implement a partnership comprised of government agencies, oil and gas pipeline operators, political representatives, and maritime representatives. The SGA, the Offshore Operators Committee, the DOT, and other related industries have started the process. The process involves the following steps to implement the one call system.

A. Establish a reliable data base on the position of offshore pipeline facilities using DGPS technology.
B. Task group to establish one call system.
C. Set up enforcement provisions of one call system.

Safe Mooring Areas:

Another part of the damage prevention offshore is to establish safe mooring areas where boats, ships, anchor handling boats, jack up's, etc., can safely moor during foul weather without worry of pipelines and environmental sensitive areas.

Offshore Public Awareness Program:

The offshore public awareness program is under way at this time through the SGA. However, the program needs to be expanded to be more encompassing. Surveys used in the one call system could be published through NOAA and used be in the Gps system or put on navigational charts. Funds from the Fisherman's Fund could be used to prevent net replacement rather than replacing nets after the fact.

Emergency Plans:

100

Discussion is currently under way between producers and pipeline companies to respond to emergencies quicker (joint partnership) through a coordinated effort.

Some of this information was discussed in the Mineral Management Service and the Office of Pipeline Safety sponsored International Workshop on Damage to Underwater Pipelines on February 22-24, 1995 at the Doubletree Hotel in New Orleans. Discussion was on shallow water surveys and prevention and remediation of shallow or exposed pipelines. The direction the government is looking at going on shallow water surveys. What is a reasonable approach to the problem of shallow pipelines. The safety alert on accidents involving rigs, barges, and anchor handling vessels. Results of the hurricane damage survey. How an offshore one call system for rig movement, line crossings, dredging, could work. Where the MMS is at one the digital conversion of the pipeline maps. Making the data base available to all surveying companies and to the one call system. I am sure we will get into many other facets of the preventing and mitigating operational problems.

Recommended by: Brian Webb, BKW, Inc.

Effective Use of Auger Anchors

Justification: Pipelines crossing swamps require negative hold down and this can be provided using concrete set-on weights or auger anchors on large diameter (36") pipelines. Auger anchors can save $80.00 per foot over concrete set-on weights. However, pipeline companies are reluctant to use auger anchors because of past failures. These failures are a result of poor engineering practices. A study could be made to indicate that with proper engineering practices the reliability of concrete can be achieved using auger anchors.

Recommended by: Mike Rickman, City of Dallas Water Utilities Department

Water Main Replacement Criteria

Justification: The determination to replace a potable water main in lieu of it's continued replacement is predicted upon a number of factors, some of which are simple and some of which are quite difficult to quantify for analysis. At the present time, there are no sophisticated software programs which will accept the input of all known data, analyze them, and recommend what action should be taken, and when it should be taken. The development and wide-spread acceptance of a standardized computer model which can analyze a main for replacement would be a boon to the industry, and would assist many utility companies in obtaining funding for expanded main replacement programs.

The program should be capable of utilizing not only the basic information related to the pipeline material, age, depth of bury, condition of the pipeline, the number, cost, and history of breaks, but also considerations related to the sensitivity of the pipeline, such as the extent of customer inconvenience when the main has to be shut down, the political sensitivity of the location of the breaks, and the degree of damage to public and private property resulting from main breaks.

Water Main Rehabilitation

Justification: The increasing urbanization and the resulting congestion found everywhere are driving factors in forcing many utility companies to consider alternative construction methods such as "trenchless" technology to minimize the impact of construction activities on the general public and upon other utility companies. Some areas, however, have become so highly congested that it is virtually impossible to replace a buried utility line of any type, by any means. More research is desperately needed into alternative rehabilitation processes for potable water lines, such as in situ lining to provide relief from a problem that only grows worse each year.

Understandably, the bulk of the research to date into in situ linings has been directed towards wastewater applications, which has paved the way for refinement of this technology for applications in potable water mains if sufficient interest is shown.

Recommended by: Mel Kanninen, MFK Consulting Services

More Widely Applicable, Accurate and Less Intrusive Inspection Methodologies

Justification: The most common methods currently used for conducting transmission pipeline NDI for structural integrity purposes are visual examination, hydrotesting, and instrumented pigging. While all of these methods have value and will undoubtedly continue to be useful to the industry, they also each have significant drawbacks. What is needed to supplement these are techniques that can assess the condition of the pipeline over long distances (i.e., 50 miles) without requiring the line to be completely shut down, nor cause extensive physical deterioration of the steel, the coating, or the surroundings. A possible way in which this could be done is by using the pipe wall, and/or the pipe and the fluid that it carries, to transmit a signal that is measurably perturbed by the presence of physical damage or degradation in the wall between the transmitter and the receiver.

Quantification of Pipeline Damage/Degradation Mechanisms

Justification: There are many different types of damage that can be inflicted upon a pipeline, by man or nature. These include wall thinning from general corrosion, possibly associated with localized coating failures; dents, surface scratches, and gouges; and stress corrosion cracking. Research is needed that systematically integrates modern computational analyses, laboratory-scale experiments, material characterization, and engineering models for industry use. In this approach, full-scale experiments are used judiciously to guide the research and to serve as independent "proof-of-concept" validations. Work of this type cold usefully be focused on all of the above-mentioned pipe damage and degradation mechanisms, as well as being linked to in-service pipeline NDI, both to set the target in advance of inspections, and to accurately quantify their findings.

Transmission Pipeline Rehabilitation Procedures

Justification: In the past few years the gas distribution, water and sewer industries, and the electric power and communication industries have embraced the benefits of "trenchless" or "no-dig" technology through sliplining, pipe bursting, and modified sliplining. Accordingly, a great many techniques and providers of those techniques now exist. One obvious problem with bringing this technology directly to transmission pipelines is that the latter have a high pressure requirement that the high density polyethylene and PVC liners used in the fold-and-reform methods, and the woven composite fabrics used in the cured-in-place methods do not offer. What appears to be needed is the identification of composite materials (fiber-reinforced polymers and/or sandwich composites) that have sufficient flexibility, durability, high strength and impermeability to utilize the proven modified sliplining technology to be inserted as liners.

Group D: Fluid Mechanics/Hydraulics of Pipelines

Includes dynamic analysis of pipeline transients, water hammer and column separation, cavitation in pumps and valves, rheology of slurry, and hydraulics of capsule flow.

Facilitator:

Benjamin E. Wylie
 Professor and Chairman
University of Michigan
Department of Civil Engineering
Room 2342, G. G. Brown Building
2305 Hayward
Ann Arbor, MI 48109
(313) 764-6499

Panelists:

Tom Walski
Professor of Civil Engineering
Wilkes University
P.O. Box 111
Wilkes Barre, PA 18711
(717) 831-4882

John Bomba
Kvaerner - R. J. Brown
1253 North Post Oak Road
Houston, TX 77055
(713) 957-5914

Steven J. Troch
Baltimore Gas and Electric Company
1699 Leadenhall Street
Baltimore, MD 21230
(410) 291-4540

Recommended by: E. B. Wylie, Department of Civil and Environmental Engineering, University of Michigan

Valve Dynamics - particularly check valves

Justification: Check valves are inserted in systems to protect system components, to prevent back flow and water hammer, among other reasons. Yet with every installation there is a concern about actual operation. Under some operating condition will the check valve slam, or will it vibrate?

The need is great for experimental data on various types of check valves and on various size valves. Simulations would then be able to be validated so they could be used reliably in a design and analysis mode. Additionally, with good experimental data in hand, dynamic similarity studies should result in the identification of the parameters most important in the selection of an appropriate valve type for a given application. This could easily precipitate the design of new-improved valve types by manufacturers.

Transient Flow Induced by Thermal Events

Justification: Although uncertainties remain in modeling cold water vaporization and condensation in pipeline systems, there are major unknowns associated with condensation-induced water hammer. The need is most critical in the nuclear power industry.

Limited data are available in laboratory-scale experiments, and these data form the basis for any theory that currently exists. Experiments are needed on larger apparatuses - up to full scale, since size scaling is not apparent. Thus, current modeling techniques do not necessarily conservatively predict the performance of prototype systems.

Leak Detection in - oil product lines,
natural gas lines, and
water distribution systems

Justification: Leak detection is necessary on some systems for safety reasons, for environmental reason, for economic reasons, among others. The oil and gas industries have invested enormous resources in leak detection systems to monitor their operating pipeline systems. Yet the "ideal" method remains illusive.

The obvious approaches have been implemented, some with great care and understanding, but not with total satisfaction. Uncertainties in data gathering, transmission, handling, and in modeling procedures plague this particular application problem.

106

Recommended by: Thomas Walski, Wilkes University

Water Quality Changes in Water Distribution Systems

Justification: Over the last decade, changes in the Safe Drinking Water Act have made it much more important than ever to understand and model changes in drinking water quality between the treatment plant and the customers' taps. Existing models do a reasonably good job in predicting the fate of conservative substances (e.g. fluoride, salinity) and substances that obey a first order decay (e.g. chlorine). There is still a great deal of work that needs to be done handing more complicated substances (e.g. pH-alkalinity-hardness or sediment transport). Many times reactions that occur at a given rate in laboratory vessels do not occur at the same rate in water distribution pipes. This is due to reactions with pipe walls and biofilms. These reactions and rates are not well understood. There needs to be some basic research identifying the kinds of films and sediments that are formed in water distribution systems and their chemistry.

Manual Practice on Trenchless Technology

Justification: Over the last decade there have been dramatic advances in trenchless (No-Dig) technology for utility pipelines. Civil Engineers are currently overwhelmed by the choices and need a basic Manual of Practice on the subject--not a collection of disjointed conference papers or advertising provided by vendors. What is needed is a document that can be used by design engineers in preliminary design stages of a project to identify the most appropriate technology.

Frost Protection for Water System Components

Justification: Freezing of water pipes has always been a problem for water utilities and their customers. Recent more stringent requirements for increasing levels of backflow prevention have increased the complexity of protecting valves and pipes from freezing in cold climates. With the large number of backflow prevention devices being added to systems every year, a small improvement in freeze protection can save millions of dollars.

Recommended by: John Bomba, Kvaerner - R. J. Brown

Calibration of Prediction Techniques for Slug Flow and Slug Length

Justification: Field data from operating multi-phase flow pipelines is required to calibrate prediction techniques for all commercially available multi-phase programs and publish correlations for slug flow and slug length.

Slugs due to pigging and terrain features are predictable within plus or minus 20 percent. However, prediction techniques for hydronamic slugs are particularly bad.

Similarly, prediction techniques for determining slug length are not good. One program has a reported "accuracy" of plus or minus 100 percent -- and this is the best of the lot. The next best has a reported "accuracy of length prediction" of plus or minus 1000 percent.

Recommended by: Steven J. Troch, Baltimore Gas and Electric Company

Ultrasonic Weld Inspection

Justification: Currently ultrasonic technology is applied for steel pipeline production seam welds (ERW process) but the inspection of circumferential field welds continues to be performed by radiography (NDE). Research could be initiated to develop a way to apply ultrasonic inspection technology to field processes. Any process developed must be applicable to pipeline construction conditions and be accepted by DOT, ASME and API as meeting NDE requirements. Such a process could have significant benefits in the elimination of radiographic sources and safety issues related to the current practice. Potential economic savings in material costs (and availability), productivity and ability to revise construction sequence practices.

New Concept for Locating Underground Plastic Pipe

Justification: Within the last several years the use of "plastic" pipe, which includes PVC, fiberglass, PE and others, has increased significantly to the point that today the majority of all piping systems installed in the US are plastic. A residual effect of this growth in plastic systems is our ability to provide accurate locating of underground piping following installation. Research efforts are needed in developing a universal technology in locating non-magnetic underground piping utilized for any medium (water, sewer, gas, air and duct). Although some efforts have been initiated by the gas industry, research needs to be expanded to encompass all applicably systems rather than focusing a proprietary solution for each medium. The economic costs of damage to underground systems due to inability to locate is growing rapidly and will continue to be the largest maintenance (and safety) cost associated with these installations.

Composite Materials

Justification: Research efforts could be promoted to expand the acceptance and use of structural composite materials and expansion of the technology into use of carbon fiber resins. This technology could be applied to use in pipeline bridge structures, sign structures, shoring, plates and underground structures such as manholes or vaults. This application would have significant benefits in corrosion deterrence, weight reduction for constructability savings, material costs and potentially developing structural systems which can be "field-adaptable".

Group E: Construction and Rehabilitation of Pipelines

Includes new construction techniques for pipelines, and construction under extreme conditions such as cold regions, mountainous terrains, swamps, and wetlands, and offshore conditions; in-situ lining, replacement of corroded pipe segments, retrofitting of existing pipelines to comply with earthquake design, renovating decommissioned oil pipelines and natural gas pipelines for other purposes such as transporting coal.

Facilitator:

B. J. Schrock
President
JSC International Engineering
1313 Gary Way
Carmichael, CA 95608
(916) 483-8170

Panelists:

Ken Kienow
President, Kienow & Associates
Inc.
P.O. Box 121110
Big Bear Lake, CA 92315
(909) 866-8636

James Baker, Jr.
President, Baker Pipeline
206 Industrial Avenue "C"
Engineers
Houma, LA 70363
(504) 868-2854

Kent A. Alms
St. Louis County Water Company
535 North New Ballas Road
St. Louis, MO 63141
(314) 997-1662

Arun K. Deb
Vice President, Roy F. Weston, Inc.
Suite 1515
1515 Market Street

Dan Cook
Vice President, American Liner,

506 Carmony lane, NE
Albuquerque, NM 87107
(505) 344-7719

Ahmad Habibian
Manager, Technical Activities
American Society of Civil

1801 Alexander bell Drive
Reston, VA 20191-6000
(703) 295-6071

Philadelphia, PA 19102
(215) 841-2014

111

Recommended by: Kenneth Kienow, Kienow Associates, Inc.

Structural Adequacy of MIC Corroded Concrete Sewers

Justification: The process and extent of microbiologically induced corrosion (MIC) in concrete pipe sewers is very dependent on localized turbulence within the pipe. Turbulent areas are usually limited in their longitudinal extent within the pipeline, and are generally limited to a few tens or hundreds of feet downstream of points of unusual hydraulic turbulence. The pipe deterioration is limited to the unwetted upper half of the pipe, above the low flow line. The severity of the corrosion is indicated by the inches of crown or springline concrete lost, the amount of original reinforcing steel remaining, and whether or not additional steel is present ("double circular cage" reinforcing steel) in the portion of the pipe remaining.

The common practice to rehabilitate several thousands of feet of pipe, and often several miles of interceptor sewers, when in fact the length that is structurally impaired is a small fraction of the total length. Rehabilitation costs run in the range of from five to ten dollars per inch of pipe diameter per linear foot, so the cost for five foot diameter sever can run about $1.5 to $3 million per mile. Rehabilitating 100 percent of a five mile long major interceptor sewer when less than ten percent is structurally impaired is a gross waste of scarce public resources.

Several factors lead to the over conservatism exercised by rehabilitation designers, not the least of which is the liability issue, since in most municipal contracts the designer is required to "hold harmless and defend" the agency for any and all future problems, regardless of fault. This is a case where the "risk management" attorneys on the owner/agency's staff increase the expenditure of tax dollars by factors of five to ten or more, thinking they are "protecting" the agency/client. The second factor, and one which requires some testing and research is the need for verification of design methods which are capable of predicting the load carrying ability of a partly corroded pipe. The fact that the pipe, prior to rehabilitation, is carrying soil and live loads successfully is certainly evidence that the corroded pipe possesses significant structural capability.

In the past, part of the justification for wholesale rehabilitation of major sewers was that the new liner, being inert, was meant to control future corrosion of the interceptor. Recent advances in corrosion control techniques, such as the magnesium hydroxide slurry sewer crown spray will arrest corrosion completely for a year or more, at which time the treatment is repeated. The annualized cost of corrosion control, according to the Los Angeles County Sanitation Districts, is far less than the cost of rehabilitation.

Simple three hinged arch analysis indicates that the pipe, with half the concrete and some of the steel gone, can carry nearly the same loads as the original pipe. There is a need for research which will enable the engineer to accurately assess the structural capability of existing corroded pipe.

Friction Factors for Rehabilitated Pipelines

Justification: Existing corroded and/or structurally or hydraulically deteriorated gravity pipelines are rehabilitated by inserting a smaller diameter pipe into the existing pipe or by constructing a liners within the pipe by various methods. The rehabilitation design engineer must be able to accurately estimate the flow quantity which can be handled by the new smaller diameter pipeline. Recent pre and post-rehabilitation data collected by the Los Angeles County Sanitation Districts have indicated extreme variability in observed friction factors and hence in the flow capacity of the pipeline. The current inability to predict capacity may result in serious environmental impacts. Research is needed which will provide data for accurately forecasting flow capacity for various types of rehabilitation options.

External Hydrostatic Long Term Buckling Resistance of Pipeline Rehabilitation Liners

Justification: Liner pipe in rehabilitated pipelines is subject to buckling from external pressures, generally due to groundwater infiltrating the joints, cracks, or other defects in the host pipe. Grouting the annular space between the liner pipe and host pipe greatly increases the allowable external hydrostatic head. Common design practice requires twenty five to fifty year life expectancy for the rehabilitated pipe. The long term ability of the lined pipe to resist buckling is a function of the material properties of the host pipe; the properties of the grouting material used; the dimensions and particularly the thickness and radial extent of the grouted/ungrouted annular space; and the material properties of the liner pipe, particularly the modulus creep characteristics. Existing test data (by TTC, Louisiana Tech) are statistically grossly erratic, and include only a narrow range of the variables.

113

Recommended by: James R. Baker, President Baker Pipeline

Deep Water Development - Gulf of Mexico

Justification: The emerging deep water development in the Gulf of Mexico is creating an opportunity for technological advancement that has not been seen since the "Space Race". This development involves new methods of pipeline construction and connection work at depths of 2000 to 6000 ft. Of water and correlating external pressure of 873 psig to 2600 psig. Casings, valves, pipes, insulation, drilling techniques, cathodic protection, and other hardware and services that we take for granted on surface work will not begin to work with these external pressures. The Gulf can and will be proving ground that will open up the rest of the world to vast reserves of oil and can.

Environmental Impacts and Government Regulations

Justification: Pipeline construction, in particular rehab construction, is facing permit restrictions, regulations and construction requirements that are not economical, practical, or sometimes even feasible with the current technology. Engineers, construction managers, and pipeline owners with the real world construction experience need to address these items with a unified voice before our industry is doomed to a slow death similar to the one experienced int he Nuclear Industry.

Trenchless Technology

Justification: The excavation of a pipeline any type of construction opens a vast area of opportunities for any casual and uninformed spectators to create problems for all parties involved. The development of directional drilling and trenchless technology has helped mitigate this problem. It will be more important to develop new ways to install pipelines without having exposure to all spectators.

Recommended by: Kent A. Alms, St. Louis County Water Company

Water Main Replacement Techniques

Justification: Aging infrastructure and impending water main failures will be an ongoing concern for Water Companies in the future. Currently, most water main is being replaced with iron pipe by conventional trenching methods. Standard construction practices that are intended to inhibit future corrosion on new pipe include, polywrapping and installation of insulating corps and couplings.

The vast majority of the older mains in need of replacement are located in developed neighborhoods with mature trees and other landscaping. Also many of the mains are located under pavement. In order to minimize disruption of service and destruction of neighborhoods, new construction techniques need to be research for application in the water industry. For example, trenchless technologies including pipe bursting, with the subsequent installation of a plastic pipe (to be used as the carrier or as casing) are methods that deserve investigation. Other technologies such as slip-lining and plastic "U"-liners should also be researched. The intent of researching the above technologies is to minimize the destruction caused by open trenching methods.

PE, PVC or Other Plastic Carrier Pipes

Justification: Currently most large water utilities install cement lined, DI polywrapped pipe as the product carrier pipe of choice. PVC and PE product carrier pipes, have also recently been used, by typically smaller water utilities. St. Louis County Water Company has two important concerns that must be addressed before they would consider using either PVC or PE as carrier pipes. The first concern revolves around the high coefficient of thermal expansion and contraction of plastic materials. For water utilities using surface supplies that vary greatly in temperature range over the course of a year, the expansion and contraction of plastic materials could lead to main and/or service line failures. The issue deals with water treatment; St. Louis County Water Company produces a scaling water, which deposits and adheres to the inside of pipelines over time. This is a good, well established treatment practice for other reasons too numerous to mention here. However, we have concerns that this scale will break free from the inside of plastic carrier pipes when undergoing expansion and contraction due to temperature changes. This may ultimately lead to clogged meters and service lines.

Corrosion Mitigation for Existing Pipes

Justification: There is a current need for research in the area of corrosion mitigation for existing pipelines. Research to develop intermediate solutions before the final and drastic measure of conventional trenching and replacement become necessary, may be beneficial in the future to water companies in combating deteriorating infrastructure. Corrosion on iron water pipes occurs due to many factors. Studies have shown that some corrosion may be caused by stray electrical currents coming from houses where the copper water line is used as a grounding device. Grounding to the water service line is still a common practice and is specified in the National Electric Code. Insulated corps and fittings can be used as a somewhat effective stop gap measure to prevent stray current on service lines.

Another source of corrosion on iron pipes, is sulfur reducing bacteria in the soil attacking leadite joints; this causes a decomposition of the joint resulting in the ultimate failure of the pipe at the joint. A possible solution to this dilemma may be drilling holes at joints and injecting a bacteria destroying solution. In summary, intermediate solutions to corrosion problems that may prove cost effective to water utilities in combating deteriorating infrastructure, should be investigated.

Recommended by: Arun K. Deb, Roy F. Weston, Inc.

Water Main Breaks Due to Water Temperature Change

Justification: In general, 80 percent of water utility investment is committed to distribution and transmission pipelines. Water pipelines in many cities in North America are old and consist primarily of unlined cast iron pipes. The lack of proper and timely maintenance of water distribution network systems is causing the incidence of main breaks to increase at a high rate in many cities. It has been observed that water systems using surface water as source are having large numbers of breaks during a short period of time when temperature of water changes rapidly.

This phenomenon has not been studied properly. A systematic study to understand causes of these breaks and to identify measures that a water utility should take to avoid or reduce number of breaks. This study result will be highly beneficial to water distribution system managers.

Water Main Renewal/Rehabilitation Program

Justification: In order to rehabilitate and renew water distribution pipelines of this country, it is estimated that approximately $100 billion is needed over the next two decades. Since all distribution system pipelines are below the surface, conditions of pipelines are generally not known. In order for water utilities to develop a cost-effective program for water pipeline rehabilitation/replacement program. There is an urgent need for development of:

• Technologies which will identify structural conditions of water pipelines.

• A predictive distribution system condition assessment model that describes deterioration process of water mains.

Recommended by: Daniel W. Cook, Cook Construction Company, Inc.

Evaluation of Long Term Physical Characteristics of HDPE & PVC Conditions

Justification: Many municipalities are experiencing some degree of premature failure of plastic distribution lines. The soil conditions, loading, and function of the line facility all contribute to these deficiencies. Excavating existing segments of lines and their subsequent laboratory testing to compare the physical characteristics of the old pipe to the ASTM standard requirements for manufacturing and design would enhance the understanding of how the environment effects the life cycle of the plastic. This would therefore assist engineers in making the decisions concerning the design characteristics of the products to use and allow proper programming of capital expenditures.

Development of Remote Pipeline Condition Assessment Equipment for Evaluating Steel, Ductile Iron and Cast Iron Pipes

Justification: The emerging technology in pipeline condition assessment by in situ electromagnetic measurements using water pressure as the transporter has been successful. Nevertheless, devising a means of transporting similar equipment through longer reaches of pipelines without the limitations of a cable or wires would significantly reduce the cost of pipeline evaluations.

Develop a Trenchless Technology Public Information/Educational Committee within ASCE to Promote the Advantages and Social Cost Savings to the Public

Justification: The cost savings of having a social proactive community effort to rehabilitate and renovate the underground infrastructure before open cut repairs and replacement are required would result in immense economical benefits to the municipality. The open cut methods of pipeline rehabilitation have always been a hotbed of homeowner and business complaints with negative public relations as a result of the disruption and damage. An educational committee that developed information and literature highlighting trenchless technology and the efforts by the owner and engineer in addressing the concerns and social impact to the public would create a partnership with all players.

118

Recommended by: Ahmad Habibian, American Society of Civil Engineers

Development of Condition Assessment Technologies for Water Mains

Justification: A key factor in rehabilitating water distribution systems is the ability to asses the condition of buried mains. Only after the condition of the system is known can an evaluation of feasible rehabilitation measures be made. The Intelligent Pigging technology has been used extensively in the oil and gas industry since the 1960; however, the water industry has been hesitant to try such techniques because of cost and operational issues. To overcome these issues, the water industry should pull their resources together and adapt the technology from the oil and gas industry. The project will identify the barriers preventing utilization of this technology in the water industry; formulate solutions to overcome the barriers; modify the existing tools to meet the needs of the water industry; and will conduct several pilot projects.

Structural Rehabilitation of Cast Iron Water Mains

Justification: There are thousands of miles of buried cast iron water mains which are deteriorating at various rates. Many utilities have developed limited rehabilitation of their systems. By far, the most widely used pipe rehabilitation technique is the cement mortar lining. The cement mortar lining is essentially a non-structural element which does not increase the pipe's structural strength. A number of synthetic materials are available which can be used for structural rehabilitation of water mains. In Europe, for example, polyethylene is used for rehabilitating old cast iron mains. The potential health effects of synthetics and their long-term performance need to be investigated, before the water industry embraces the use of such materials for rehabilitating water mains.

Earthquake Hazard Effects

Justification: Damage to distribution systems during earthquakes is a serious concern in certain parts of the country. Fault movements and soil liquefaction are two primary causes of pipe failures due to such events. Also, disruption of a water supply significantly impairs the ability of fire fighters to extinguish fires which may start after an earthquake. In the 1906 San Francisco disaster, fire damage was much more severe than that caused by the earthquake itself. Further research in design, installation, and performance evaluation of pipelines under earthquake loading is necessary.

119

Group F: Automatic Control and Instrumentation

Includes computers and other new technologies used for automatic control of pipelines, control strategies, and communication systems; includes measurement of flow, pressure, and temperature of the fluid in the pipe.

Facilitator:
William Hunt
MSE-HKM Engineering
P.O. Box 1090
Bozeman, MT 59771
(406) 586-8834

Panelists:
Aubrey F. Zey, President
Nova Tech, Inc.
13604 West 107th Street
Lenexa, KS 66215
(913) 451-1880

Edward J. Farmer
EFA Technologies
1611 20th Street
Sacramento, CA 95814
(916) 443-8842

Roy Fleet, Senior Engineer, Health & Safety
Natural Gas Pipeline Company of America
112 N Lincoln
Westmont, IL 60559
(708) 691-3786

William F. Quinn, Manager
Codes and Standards
El Paso Natural Gas Company
P.O. Box 1492
El Paso, TX 79978
(915) 541-5121

Don Scott
Interprovincial Pipe Line Co.
10201 Jasper Avenue
Edmonton, AB T5J 2J9
(403) 420-8118

Recommended by: William A. Hunt, MSE-HKM Engineering

Ultrasonic Meters for Gases and Liquids

Justification: Recent developments in digital electronics, signal processing and enhanced software techniques have improved the accuracy and reliability of ultrasonic meters for measurement for transfer of custody. As these devices are non-intrusive and not affected by the build-up of biofilms and precipitates, have bi-directional flow capability, require no pipeline restriction and are unaffected by pulsating flows, they offer many advantages. The current state-of-the-product requires extensive beta testing, establishing standards for testing procedures, development of standards of accuracy for configuration of single and multi-path meters and analysis of electromagnetic interference on the signals and telecommunications links to central data processing. The use of ultrasonic meters in conjunction with programmable logic controllers (PLCs) for distributed control systems (DSC) needs further research and development.

Leak Detection Processing from SCADA Systems

Justification: The reduction of revenue from leakage and the increased concern about contamination of groundwater accentuate the need for better leak detection. Continuous processing of SCADA data using the methods and software developed offer improvement of leak detection. Three issues need to be addressed: (1) frequency of scan, (2) computational methods and software to correct skew of data received from different sensor locations, and (3) additional beta testing of complete systems from analog sensors through to the reading of the results to check for leaks and accuracy of predicting leakage rates and locations.

Formulation of Program Logic Controllers (PLC) Responses to Fluid Transients

Justification: Over the past 30 years computer simulations have been developed for increasingly complex upsets causing potentially destructive fluid transients in pipeline systems. The effect of remedial measures on the flow characteristics is a function of (a) the time profile and (b) the variation of the geometric configuration of the controlling element during the time which the controller operates. The controller action generated by the response entered into the PLC produces flow characteristics which will effectively neutralize the harmful effect of the transient. The effect of the movement of the controlling element is largely unknown. Destructive transients in pipelines with transient control devices have been noted. Research to develop a body of data on the controller operation on flow characteristics for different devices is needed.

Recommended by: Aubrey F. Zey, NovaTech, L.L.C.

Leak Detection

Justification: There are two types of pipeline leak detection systems avaiiable today: sophisticated systems that use dedicated computers, and simple ones that run on a PC

The sophisticated leak detection systems often require dedicated personnel to evaluate the data and make periodic adjustments. The added expense for these types of systems do not provide results that are much better than SCADA operators monitoring pipeline operations. The PC versions do not yield results any better than SCADA operators monitoring pipeline operation.

The better results are for short lines running the same product. There is a specific need for leak detection methods for long pipelines with multiple products.

Power Optimization

Justification: There is a need for an on-line power optimization model. The mathematical model would accept booster station inputs with regard to unit configuration and associated variables and local energy price structure. The model would optimize booster station pumping configuration for optimum revenue.

Electronic Data Interchange

Justification: The current Electronic Data Interchange service transfers product delivery information and invoices electronically. The information this system relies on usually requires some manual input. The current system provides acknowledgments of data transfer and security.

There is a desire to reduce the time between product delivery and submittal of invoice. Ideally, the system would generate an invoice as soon as the product has been delivered, automatically without any manual input.

Developing a process using the Internet has been suggested. A standard protocol for Electronic Data Interchange would be required to implement this type of system. The protocol would include, but not be limited to, the data format, security and procedures.

Recommended by: Ed Farmer, EFA Technologies, Inc.

Emergency Response Plan Implementation

Justification: Regulations require that pipeline operators have an emergency response plan for which there are detailed requirements. The criteria that trigger implementation of the plan are not identified in the same detail, if at all.

Research is need to establish the optimal way in which to trigger an emergency response in accordance with an approved plan. This research should directly address the methodology to be used for detection of an accident. It should assess the costs of accident detection as well as the benefits stemming from faster action resulting in smaller discharges and consequently less environmental damage and lower clean-up and mitigation costs.

Automatic Acting and Remote - Controlled Line Block Valves

Justification: Regulations and permit conditions increasingly require automatic and/or remote controlled line block valves. These are intended to quickly sectionalize a pipeline in the event of an accident and thereby minimize environmental damage.

From a system safety and reliability point of view, line block valves are not free of risk. They are less reliable than line pipe hence installing them can increase accident probability. They can also produce hydraulic hazards due to surge if they are not properly designed and implemented. (In some cases permit conditions have required dangerously rapid valve closing rates.)

Research in risk analysis is needed to provide a clear assessment of the safety impact of line block valves. A product of this research could be a standard method for assessing the impact on system safety and reliability on a valve-by-valve basis.

Value of Training Simulators

Justification: There is considerable interest by industry and regulators in pipeline simulators for training operators; however, there are no clear criteria regarding what these simulators should be capable of or how well they should simulate actual pipeline operation. A poor training tool may actually diminish the performance of the trainees.

Research is needed to identify proper training goals and needed simulation quality, and to develop methods for auditing the effectiveness of simulator-based training.

123

Recommended by: F. Roy Fleet, Natural Gas Pipeline

Improvements to Line Break Detectors

Justification: Recent research published by the Gas Research Institute documents the line break detection capabilities of current automatic and remote control valves installed in natural gas transmission pipelines. The reliability of current detection methods is limited by S/N ratio. Further research could result in novel sensors or detection methods for use in natural gas transmission pipelines.

Recommended by: William F. Quinn, El Paso Natural Gas Company

Automatic Control Valves (ACV) and Remotely Controlled Valves (RCV) for Mainline Gas Transmission Pipelines

Justification: Present equipment used by the natural gas transmission industry for detection and control of pipeline breaks has proven unreliable for many applications. While the pipeline valves and their gas/hydraulic operators normally perform adequately, the detection systems and logic control used to trigger the closure of automatic valves are plagued by reliability problems. Most detectors seek to identify a rupture event by monitoring transient pressure signals that are generated in the pipeline by the quick release of gas. Further line break control improvement could result from research on novel sensors that can better discriminate when a pipeline break occurs. Also the continued development of "intelligent" or "smart" pipeline valves with distributed measurements and control features, in conjunction with increased pipeline automation and telecommunications capabilities, will result in future reliability improvements. (GRI Report No. GRI-95/0101)

Excess Flow Valves (EFV) for Gas Distribution Service Lines

Justification: For the last several years there has been a technological controversy regarding the application and effectiveness of Excess Flow Valves (EFV) for use on natural gas distribution and residential service lines. The US DOT/OPS, the gas distribution industry, the gas industry associations, and many gas research agencies have disagreed with the technical effectiveness and cost-benefit analyses which have been promoted by the National Transportation Safety Board (NTSB) and the various manufacturers of EFVs. The Manufacturers Standardization Society (MSS) Standard Practice SP-115 prescribes certain minimum standards for EFVs used in natural gas service. This standard applies to any EFV intended for use in a gas service line with a minimum operating pressure no less than 5 psig and specifies allowable tolerances for "bleed-by" or leakage around the EFV. The American Society for Testing and Materials (ASTM) is developing a recommended practice (RP) for the testing of EFVs and a specification establishing performance standards for EFVs. The current state-of-the-product (EFVs) and the development of testing and performance standards certainly warrants further field validation testing and technology transfer that could be gained through applied research conducted jointly between the EFV manufacturers and the gas utility users groups.

Remote Monitoring and Control of Pipeline Systems

Justification: The recent developments in the technology of electronics and communications are just beginning to have wide-spread applications in the gas pipeline industry. The following listed developments and applications are at various stages of commercial implementation, all of which could benefit by applied research through further field testing and technology transfer:
* Remote Sensing for Cathodic Control Systems - Magnetometry Surveys and Communication Relays (for the purpose of acquiring and transmitting date).

- Gas Leak Detection - Infrared or Thermal Imaging Photography/Video and Software-Based technology (using volume balance or rate-of-change methods).

- Monitoring Rights-of-Way for Third-Party Encroachments and Environmental Alterations.

All of these above listed remote monitoring techniques should be developed utilizing both airborne (aircraft or helicopters) and global satellite technologies.

Recommended by: Don Scott, Interprovincial Pipe Line Co.

Application of Expert Systems

Justification: Almost all pipelines are highly computerized. They employ sophisticated SCADA systems which remotely control the pipeline from a central location and generate huge volumes of data. Other than for some preset limit values, control of pipeline is completely done by people (pipeline controllers). Also, the SCADA data is normally used (viewed) just once in scan then archived never to be considered again.

Most pipelines have been operated for a number of years so a great deal of expertise resides in the control centres. Because of shiftwork different pipeline controllers operate the line - this leads to a variation in operation some of which may not be optimal. It may be possible to capture "best" operating practices, build an expert system for automatic control or at least initially to provide suggestion to a pipeline controller and therefore increase the efficient of operation. Also the SCADA dataset may be able to be used in an expert system for other applications - such as monitoring the calibration of the instruments. It would appear that pipeline may be an ideal application for expert systems.

Instrumentation

Justification: Almost all pipeline functions are remotely controlled so a pipeline uses a great number of sensors to provide an indication of pressure, temperature, flow and so on. All these instruments are intrusive, so this makes installation expensive and provides a potential leak source.

Non-intrusive, but accurate instruments to measure pressure, temperature, flow, density, and viscosity, for example wold be welcome in the pipeline industry. Some non-intrusive instrument that already exist, for example sonic flow meters, cannot yet be used for highly accurate measurement of flow. Improvements in this technology would make measurement simpler, less expensive to install and less prone to leakage. For those companies who do extensive computer modeling, real time measurement of viscosity is necessary, yet no rugged, non-intrusive instrument exists.

127

Group G: Freight Pipelines

Includes slurry pipelines, pneumatic pipelines, capsule pipelines, and tube transportation systems.

Facilitator:

Henry Liu, Director,
Capsule Pipeline Research Center
E2421 Engineering Building East
Columbia, MO 65211
(573) 882-2779

Panelists:

Bill Vandersteel
Ampower Corporation
P.O. Box 417
Alpine, NJ 07620
(201) 768-6014

Lawrence Vance
US Department of Transportation
Volpe Center
DTS-56
55 Broadway
Kendall Square
Cambridge, MA 02142
(617) 494-2273

Sean Plasynski
Program Director, University Coal Research
Pittsburgh Energy Technology Center
P.O. Box 10940, MS-922-300
Pittsburgh, PA 15236
(412) 892-4867

David Kao
492 Town Engineering Building
Department of Civil and Construction Engineering
Iowa State University
Ames, IA 50011
(515) 294-5933

Tom Pasko
Federal Highway Adminstration
Turner-Fairbank Highway Research Center
6300 Georgetown Pike
McLean, VA 22101-2296
(703) 285-2679

Recommended by: Henry Liu, Capsule Pipeline Research Center

Coal Log Pipeline

Justification: Coal log pipeline (CLP) is an emerging technology in pipeline for transporting coal in log (cylinder) for, using water as the carrier. CLP uses approximately one-third of the water needed by coal slurry pipeline to transport the same amount of coal, and it transports twice the amount of coal transported by coal slurry pipeline of the same diameter. Economic analysis of CLP shows that coal can be transported by CLP at a cost less than that by rail and truck under many conditions. CLP appears to be a promising new technology for coal transportation.

Extensive R&D in CLP has been conducted in the last five years at Capsule Pipeline Research Center, University of Missouri-Columbia. The work is sponsored by the National Science Foundation , US Department of Energy, Electric Power Research Institute, State of Missouri and consortium of pipeline, coal and electric utility companies. In spite of this extensive study, continued research in CLP is needed for at least two to three more years before the first commercial CLP can be demonstrated successfully. Even after successful commercial demonstration, continued R&D will be needed for many years to improve and perfect the CLP system design and operations. This calls for a long-term research program in CLP that should be sponsored jointly by government and industry. The research is interdisciplinary and includes not only hydraulics of coal log flow, but also coal log compaction, automatic control of coal log pipeline, drag reduction in coal log flow, wear of coal logs and pipes, and legal research on coal pipelines.

Hydraulic Capsule Pipelines (HCP)

Justification: Coal log pipeline (CLP) is a particular type of hydraulic capsule pipeline (HCP). Much of the knowledge gained in CLP is applicable to transporting other minerals and solid wastes by pipeline, again in log form. Moreover, by encapsulating bulk materials such as grain or fertilizer in plastic or metallic capsules or containers, many other bulk materials can be transported by HCP in situations where they are economical. Therefore, research is needed on HCP for applications to materials other than coal. Needed research includes compaction (when materials must be compacted into logs), encapsulation (when materials are to be encapsulated) and handling of capsules at terminals.

130

Pneumatic Capsule Pipe (PCP)

Justification: PCP, which uses air to move capsules in pipe, is the most promising type of capsule pipeline for intercity transportation of freight. In the ICETEA legislation, the Congress asked DOT secretary to investigate this future underground freight transportation system and report back to the Congress. Now that FHWA has completed the investigation with positive finding, DOT should start and R&D program in this area. Government support of this area of pipeline research appears most appropriate.

Recommended by: William Vandersteel, Ampower Corporation

Tube Freight

Justification: Surface transportation in the United States is at a crossroads and our highly prized mobility is increasingly threatened. Many of the Nation's roads are clogged and congestion continues to worsen. The conventional approach of building more roads has ceased to be effective in most areas of the country for both fiscal and environmental reasons.

Transportation infrastructure has the dual function of providing for the movement of people and the transportation of freight. To the extent that these two modes operate independently, the efficiency of traffic is improved. Moving people requires flexibility, convenience and speed; transporting freight requires cost-effectiveness, on-time delivery and security in transit and, because these constraints differ, they perform best when operating independently of each other.

The transportation of freight lends itself to automation, as is already done within a modern warehouse. One means to accomplish this is to "pump" closely fitting wheeled cargo carrying vehicles, called "capsules", through pipelines, generally referred to as "tube freight". The ISTEA Bill [1] provided for the US DOT to conduct a study to evaluate the feasibility, costs and benefits of pneumatic capsule pipelines for the movement of freight which led to a 1994 study [2] by the Volpe National Transportation Systems Center. In essence, the study concludes that tube freight is technically feasible and within the state-of-the-art, meriting further study.

A proposed 2 meter (6.5 ft) internal diameter tube freight system can accommodate well over 90% of all the freight now hauled by trucks. To the extent that trucks are displaced, highway safety is improved, traffic congestion reduced, damage to roadbeds, overpasses and bridges lessened, and oil consumption and exhaust pollution markedly lowered. That these are important issues is made obvious when one considers that, in 1993, trucks accounted for nearly 4000 fatalities and untold number on injuries and property damage, resulting from 200,000 police reported accidents, while trucks account for 97% of all damage to roads, highways and bridges [3]. Currently trucks consume 24% of all the oil used in our country [4]. Tube freight is electrically powered and uses only about one quarter of the energy needed for trucks. Because trucking is expensive, preliminary studies suggest that tube freight is more economical than trucks, including recovering the infrastructure cost in reasonable time.

Because of the real potential for tube freight to solve a host of pressing problems facing our transportation infrastructure today, for which no other solutions are in sight, and, taking account of the safety, environmental, economic and energy advantages of this technology, there is an urgent need for funds to implement tube freight studies, a necessary precursor to the construction and testing of a full scale prototype system.

[1] "Intermodal Surface Transportation Efficiency Act of 1991, Title VI, Section 6020
[2] "Tube Transportation" - US Department of Transportation - RSPA/VNTSC-SS-HW495-01

[3] US Department of Transportation - National Transportation Statistics - Sept. 1993
[4] "Transportation in America" - ENO Transportation Foundation - 1995 Edition - Pgs 20 & 56

Recommended by: Lawrence Vance, US Department of Transportation

Economic Feasibility of Pneumatic Capsule Pipeline

Justification: The primary research need int he area of pneumatic capsule pipelines is to assess the economic feasibility of such systems in suitable market niches. Unless such a system has some possibility of operating profitably, research into technical areas necessary for engineering development are unwarranted. Prior work sponsored by the US Department of Transportation and performed by the University of Pennsylvania indicated that such systems could be competitive for general merchandise freight in long haul markets. This work was based on costs 30 years ago. Since then cost reductions from moving freight in containers on unit trains have eliminated any competitive advantage of pneumatic capsule pipelines. New assessments of economic feasibility need to be made. Currently, markets for short and medium hauls in high land value - high congestion areas would appear to ve likely application areas for pneumatic capsule pipelines.

Terminal Design

Justification: Pneumatic capsule pipelines are inherently high capital investment systems. To be economically viable they need to maintain high throughput to keep unit costs down. A critical element in high throughput is terminal design, particularly for point to point applications where there may be as few as two terminals. Terminal design concepts need to be examined to determine their general costs and the proportion of total system costs they represent. Concepts for capsule loading and offloading should be examined as part of the terminal studies.

Market Analysis for Pneumatic Capsule Pipelines

Justification: A market analysis for pneumatic capsule pipelines needs to be made to identify likely customers for this service and their potential use a s function of unit price. Several location specific scenarios need to be considered.

Recommended by: Sean Plasynski, US Department of Energy

Pneumatic Transport (Dilute and Dense Phase Solids Transport)

Justification: Large amounts of coal and other solids are transported and processed each year, and significant amounts of money and energy are spent for these operations. Coal and other solid materials such as ash, sorbents, and catalysts are handled in many plants producing or using coal-based fuels (including coal preparation plants, industrial and electric utility plants, and liquefaction or other processing plants). Knowledge of these flows is less mature than that of single fluid flow, and this lack of knowledge limits the effectiveness in transporting and handling bulk solid materials. New developments in solid transport technologies, including measuring and monitoring devices and control technologies, will help elucidate the basic characteristics and dynamics of these flows, improve the existing methods of handling solid materials, and reduce transport and processing cost. Research can be generic to apply to other solids as well as coal and cover various mass loadings to be applicable to different processes (i.e., pulverized coal flow that may require a dense or dilute distribution depending on the use). Research should be aimed at developing novel, industrial-style instruments and devices that can diagnose, monitor, or control the flows (parameters of interest should include, but not be limited to, mass flow rate, solid concentration and velocity, and agglomeration of solid particles).

Coal Log Pipeline

Justification: Coal log pipeline (CLP) technology is a promising new technology for coal transportation, as described by Henry Liu in his recommendations. It has been extensively researched in the past several years at the University of Missouri-Columbia. In spite of this effort, there are still several issues that need to be researched and resolved before this technology is completely ready for commercialization. Dr. Liu provided a list of research needs. Several of these needs are considered to be priority issues that need to be resolved to determine the feasibility of this technology. These include the capability to produce a coal log compaction device that can produce adequate coal logs in a cost effective manner, water usage issues (especially in the Midwest), and legal issues regarding water rights and that of crossing land owned by the railroads.

Recommended by: David T. Kao, Iowa State University

Energy Efficient Capsule Freight Pipeline Transport System Development

Justification: As the global economic development continues, demand on freight transportation will increase. To build large additional or entirely new capacity of conventional transportation systems using rail, highway, air, and water may not be feasible especially in developing counties due to capital investment restriction. Under this circumstance transport through pipeline systems can become a possible alternative mode for freight delivery. Capsule transport system in particular offers the needed versatility that can be used to deliver variety of commodities in the same system. However, to make such system more attractive for use in developing as well as in developed regions, the energy efficiency of capsule pipeline systems must be further improved.

Hydraulic Capsule Pipeline Transport System as an Integral Component of Freight Transport Network in Developing Regions

Justification: In general, highest frequency of freight transportation occurs over a hauling distance of approximately 150 to 200 km range. This is a suitable distance for hydraulic capsule transport systems. It is therefore important to explore the use of hydraulic capsule pipeline as a collection and distribution limbs for a freight transportation network linking vicinity communities to a hub of rail or water transport lines in the same way as trucked containers serving similar hubs. This is particularly important for use in developing regions of the world where existing rail, roadway, and shipping lines reach only major cities.

Development of Techniques for In Situ Pipeline Transport Infrastructure Failure Detection and Rehabilitation

Justification: The nation's freight pipeline infrastructure has reached a point of needing significant attention. Development of non-destructive failure detection and in situ rehabilitation techniques will be needed and beneficial.

Recommended by: Thomas J. Pasko, Federal Highway Administration

Freight Movement in Tunnels

Justification: Tokyo has a population of 25 million and has started a program to move freight in tunnels. The US cities will also reach that size sometime in the future and we need to start planning transportation systems that can safely move freight into and out of the city expeditiously. Presently, congestions has severely curtailed truck movements in our large cities. Mixes of autos and trucks in a stream of traffic creates numerous accidents and reduces efficiency.

The railroads could move the future traffic if they could overcome the environmental hurdles, weather, and security problems, elimination of passenger traffic, and if they could automate. As an alternative, a system of automated, self-powered capsules in sic foot diameter tubes offers a possible solution to future freight movements. A basic need is to get a section of test tunnel about a mile in length that could be used to test and improve the technologies that for the most part exist today. A better test would be a tunnel segment in a very congested corridor that could be used also as a bypass freight carrier. It could be in a congested part of New York City, or as a priority mail carrier between Washington, DC and Baltimore. Building on the R&D needs beyond a test section are details on:

a. Pneumatic effects between capsules
b. Capsule configuration/automated loading
c. Switching/guiding (possibly using Denor's luggage system)
d. Power system (Linear Induction motors?)
e. Wheels/mag. lev.
f. Tunneling technologies
g. Intermodal transfers

Many of these items can be presently "brute-forced" designed, but supplemental R&D is needed to resolve questions, evaluate performance, and improve efficiencies.

APPENDIX B

PARTICIPANTS

Kent A. Alms
St. Louis County Water Company
535 North New Ballas Road
St. Louis, MO 63141
(314) 997-1662

Alex Alvarado
MMS
1201 Elmwood Park Boulevard
Mail Stop 5232
New Orleans, LA 70123
(504) 736-2547

James Baker, Jr., President
Baker Pipeline
206 Industrial Avenue "C"
Houma, LA 70363
(504) 868-2854

John G. Bomba, Chief Engineer
Kvaerner - R. J. Brown
1253 North Post Oak Road
Houston, TX 77055
(713) 957-5914

Richard W. Bonds
Ductile Iron Pipe
Research Association
245 Riverchase Parkway
Birmingham, AL 35244
(205) 988-9870

Daniel W. Cook
Cook Construction Company, Inc.
506 Carmony Lane, NE
Albuquerque, NM 87107
(505) 344-7719

Bud Danenberger
Minerals Management Service
Engineering and Technology Division
381 Elden Street, Mail Stop 4700
Herndon, VA 22070-4817

(703) 787-1559

Arun K. Deb
Vice President
Roy F. Weston, Inc.
Suite 1515
1515 Market Street
Philadelphia, PA 19102

Cesar DeLeon
US Department of Transportation
Office of Pipeline Safety Research
400 Seventh Street, SW
Washington, DC 20590
(202) 366-4595

Robert Eiber
Pipeline Consultant
4062 Fairfax Drive
Columbus, OH 43220
(216) 538-0347

John Elwood
Foothills Pipe Lines Co.
3100 - 707 Eighth Avenue
Calgary, Alberta T2P 3W8
(403) 294-4137

Edward J. Farmer, President
EFA Technologies
1611 20th Street
Sacramento, CA 95814
(916) 443-8842

Roy Fleet, Senior Engineer
Health & Safety
Natural Gas Pipeline Company
of America
112 N Lincoln
Westmont, IL 60559
(708) 691-3786

Ahmad Habibian
Manager, Technical Activities

American Society of Civil Engineers
1801 Alexander Bell Dr.
Reston, VA 20191-4400
(703) 295-6071

Tom Hoelscher
Transco
P.O. Box 7707
Charlottesville, VA 22906
(804) 973-4384

William Hunt
MSE-HKM Engineering
P.O. Box 1090
Bozeman, MT 59771
(406) 586-8834

Mel Kanninen
MFK Consulting Services
7322 Ashton Place
San Antonio, TX 78229
(210) 349-9882

Ken Kienow, President,
Kienow & Associates
P.O. Box 121110
Big Bear Lake, CA 92315
(909) 866-8636

Ibrahim Konuk
National Energy Board of Canada
5th Floor 311 - 6th Avenue
Calgary, Alberta
T2P 3H2
(403) 292-6911

Jim Liou
Department of Civil Engineering
University of Idaho
Moscow, ID 83843
(208) 885-6782

Henry Liu, Director,
Capsule Pipeline Research Center

E2421 Engineering Building East
Columbia, MO 65211
(573) 882-2779

J. R. Lehman
Trunkline Gas Company
P.O. Box 1642
Houston, TX 77251
(318) 836-5689

John McCarthy
Director of Engineering
National Energy Board of Canada
Calgary, Alberta, Canada

Wesley B. McGehee
Pipeline Engineering Consultant
14405 Walter Road, Suite 351
Houston, TX 77014
(713) 893-3080

Maher Nessim
Centre for Engineering Research
200 Carl Clark Road
Edmonton, Alberta
Canada T6N 1E2
(403) 450-3300

Sean Plasynski, Program Director,
University Coal Research
Pittsburgh Energy Technology Center
P.O. Box 10940, MS-922-300
Pittsburgh, PA 15236
(412) 892-4867

Thomas J. Pasko, Jr.
Director
Office of Advanced Research
Federal Highway Administration
Turner-Fairbank Highway Research Center
6300 Georgetown Pike
McLean, VA 22101-2296
(703) 285-2679

William F. Quinn
Manager, Codes and Standards
El Paso Natural Gas Company
P.O. Box 1492
El Paso, TX 79978
(915) 541-5121

Dale Reid
Exxon Production Research Company
P.O. Box 2189
Houston, TX 77252
(713) 966-6174

Mike Rickman
City of Dallas
City Hall 4A North
1500 Marilla
Dallas, TX 75201
(214) 670-8007

B. J. Schrock, President
JSC International Engineering
1313 Gary Way
Carmichael, CA 95608
(916) 483-8170

Don Scott
Interprovincial Pipeline Co.
P.O. Box 398
10201 Jasper Aavenue
Edmonton, AB T5J 2J9
(403) 420-8118

Charles Smith, Research Program Manager
Minerals Management Service
381 Elden Street
Mail Stop 4700
Herndon, VA 22070

Tom Steinbauer
Gas Research Institute
8600 West Bryn Mawr Avenue
Chicago, IL 60631
(312) 399-8100

Raymond Sterling
Trenchless Technology Center
Louisiana Tech University
P.O. Box 10348
Ruston, LA 71272

Harry Stewart
School of Civil and
Environmental Engineering
Cornell University
Ithaca, NY 14853
(607) 255-4734

Henry Topf, Jr.
Miller Pipeline Corporation
P.O. Box 34141
Indianapolis, IN 46234

Steven J. Troch
Baltimore Gas and Electric Company
1699 Leadenhall Street
Baltimore, MD 21230
(410) 291-4540

Lawrence Vance
US Department of Transportation
Volpe Center DTS-56
55 Broadway,
Kendall Square
Cambridge, MA 02142
(617) 494-2273

Bill Vandersteel
Ampower Corporation
P.O. Box 417
Alpine, NJ 07620
(201) 768-6014

Tom Walski
Professor of Civil Engineering
Wilkes University
P.O. Box 111
Wilkes Barre, PA 18711
(717) 831-4882

Brian C. Webb, President,
BKW, Inc.
P.O. Box 581611
Tulsa, OK 74158
(918) 584-4402

147

Theodore L. Willke
Gas Research Institute (GRI)
8600 West Bryn Mawr Avenue
Chicago, IL 60631
(312) 399-8100

Benjamin E. Wylie, Professor and Chairman
Department of Civil Engineering
University of Michigan
Room 2342, G. G. Brown Building
2305 Hayward
Ann Arbor, MI 48109
(313) 764-6499

Aubrey F. Zey, President
Nova Tech, Inc.
13604 West 107th Street
Lenexa, KS 66215
(913) 451-1880

INDEX